THE RACE NOBODY TOLD YOU ABOUT

How Quantum Computing Will Separate Winners from Losers – And Which Side Will You Be On?

DR. MITHUN MANALIKANDY, FIE

ISBN 979-8-90392-1-997

DEDICATION

For Minnu,

My four-legged daughter, whose pure, joyful spirit is the most comforting form of "entanglement". She reminds me daily that the best things in life can't be measured or calculated—just loved.

And for Appu and Nandu,

Who ask "Why?" about everything.

This book began because of your questions. I wrote it so you would never feel that curiosity is inconvenient, or that answers belong only to experts. I wrote it so you would know that "I don't know yet" is not a weakness—it is the beginning of understanding.

The universe is stranger and more beautiful than anyone first tells you.

Go figure it out.

Contents

GOAL

Why I wrote this book and why you don't need a PhD to read it

11:30 at Night

The kind of quiet that makes even your thoughts sound loud. I was in my home office, surrounded by papers, a half-finished notebook, and at least two cups of coffee that had long ago surrendered their warmth. On my screen was a research paper that opened with the line:

"Let ψ be a Hilbert space of non-trivial superposition eigenstates undergoing unitary evolution..."

I read it carefully. Then I read it again, slower this time, as though the words might rearrange themselves into something friendlier if I stared long enough.

They didn't.

I clicked to the next paper:

"The quantum state $|\psi\rangle$ *exists in a tensor product Hilbert space* $\mathcal{H} = \bigotimes_i \mathcal{H}_i$ *where decoherence rates scale as* $\Gamma \propto 1/T_2$*..."*

And another:

"Non-Abelian anyonic excitations in topological quantum systems exhibit braiding statistics fundamentally distinct from conventional fermions..."

Three hours. Three cups of cold coffee. One hundred and sixty-seven words that might as well have been written in Klingon.

Somewhere in the world, a small team of engineers was quietly testing a machine that might break the encryption protecting every bank account on Earth. And I realized something uncomfortable: the future was already being built—and most of us had no idea how it worked.

What frustrated me wasn't the difficulty. I've built large computational models. I've debugged systems at 3 AM. I'm not afraid of complex.

What frustrated me was realizing I'd been locked out on purpose.

Not because scientists wanted it that way. But because the language of the field gradually became so specialized that outsiders stopped understanding it. Behind those sentences was something powerful enough to reshape computation itself. Governments were pouring $60 billion into it. Tech giants were building machines colder than outer space. Security agencies were quietly panicking about the day encryption would break.

The revolution was already underway.

And I was locked out.

Not because I lacked intelligence. Not because I lacked education. But because somewhere along the way, quantum computing had been wrapped in language that felt designed to keep people out.

Here's what made it worse:

While I was struggling with that paper, someone in China had already launched a quantum satellite called Micius, transmitting eavesdrop-proof communications across 1,200 kilometres of open sky.

While I was Googling "What is superposition?", IBM was letting high-school students run quantum algorithms on real quantum computers for free.

And I realized something:

If a technology has the potential to reshape cybersecurity, artificial intelligence, drug discovery, and even geopolitical balance—then understanding it should not require three degrees and a tolerance for impenetrable prose.

It should require curiosity.

That's when I made a decision.

I'm going to write the explanation I wished existed. Not dumbed down. Not mystified. Just clear, honest, and human.

Here's what you're going to get:

We begin not with equations, but with something almost embarrassingly ordinary: a pebble in a clay bowl. The simplest counting tool a shepherd could devise. And from that pebble, we'll trace an unbroken line through two thousand years of human ingenuity, all the way to a machine cooler than outer space.

When we talk about superposition, I won't hand you a wave function and wish you luck. I'll ask you to imagine a spinning coin—neither heads nor tails until it lands.

When we discuss entanglement, we won't begin with equations. We'll begin with dice.

Imagine rolling one dice in New York while a friend rolls another in Tokyo. Every single time, the numbers are perfectly coordinated. You get a four, they get a three. You get a six, they get a one. Not because the dice were rigged. Not because someone is signalling across the planet. But because, at a fundamental level, certain outcomes are not decided until the moment they are measured.

That idea—the timing of when reality makes up its mind—is not just philosophically unsettling.

It's computationally powerful.

You will meet Einstein, who famously disliked quantum mechanics and spent decades trying to prove nature must surely be more sensible than this.

You will encounter Schrödinger's cat, which was never meant to be a party trick but a protest—a way of saying, "Surely this can't be right."

And you'll discover that quantum mechanics survived those protests not because it was fashionable, but because experiment after experiment confirmed its strange predictions.

And yes, you will do something most readers never expect:

You will write real quantum code.

Not toy examples on a whiteboard—real circuits, running on IBM's quantum computers, from your laptop. You don't need to be a programmer. I'll walk you through every line. And I promise you:

There is something transformative about watching a qubit respond to instructions you wrote.

So who is this for? If you keep hearing "AI" and "quantum" in meetings and suspect that nodding confidently is no longer a sustainable strategy—you're in the right place.

If you're tired of feeling like the future is being discussed in a language you don't speak—you're in the right place.

If you simply want to understand what all the excitement—and the anxiety—is about before it's too late—you're in the right place.

Here is the simplest way I can put it:

Classical computing is a straight highway: predictable lanes, clear rules, traffic moving in orderly fashion.

Quantum computing is more like a choreographed dance where possibilities overlap and interfere, where patterns emerge from probabilities, and where the rules are written at the atomic scale.

The difference isn't just speed.

It's a fundamentally different way of thinking about information itself. Like all powerful ideas, it becomes far less intimidating once someone takes the time to walk you through it.

That is what I intend to do.

Let's begin—not with equations, but with a shepherd, seventy-three sheep, and a clay bowl of pebbles.

oOo

Acknowledgements

Writing a book about quantum computing is a bit like trying to juggle atoms. By chapter three, half your ideas are in superposition—simultaneously "almost finished" and "nowhere near ready." So before we go any further, let me collapse this particular wavefunction and thank the people who made this improbable project possible.

My first thanks go to **Dr. Adam Timmons**, my former boss at Aptiv, who casually suggested I look into quantum computing. I suspect this was his strategic way of redirecting my complaints about Black-Scholes equations. Whatever the motivation, it worked. That small nudge changed the trajectory of this book.

To **YJ Babu**, our Chief Engineer, who once asked me to explain quantum computing in a staff meeting before I fully understood it myself. Nothing accelerates learning quite like the terror of public confusion.

To **Atul Varghese**, who patiently transformed my messy algorithms into something that actually ran. You made the work cleaner, sharper, and far more respectable than it had any right to be. I owe you more than a few Hyderabadi biryanis.

This book also stands on the shoulders of quantum pioneers who made all of this possible. **Feynman, Richard** reminded us that physics could be playful without losing rigor. **Erwin Schrödinger** gave us a cat that continues to haunt philosophy classrooms—though he invented it as a complaint, not a mascot. **Albert Einstein** doubted quantum mechanics so thoroughly that his objections helped sharpen it; science advances on the quality of its opposition. **Alan Turing** showed the world what computation could mean. **David Deutsch** imagined quantum computers when they were still considered fantasy. **John Preskill** gave us language—"quantum supremacy," "quantum advantage"—for milestones that once felt impossible to even name. And **Peter Shor**, whose algorithm turned quantum computing from curiosity into a cryptographic earthquake.

They taught me that revolutionary ideas don't require incomprehensible language—just someone willing to translate.

To the teams at **IBM, Google, D-Wave, IonQ, Rigetti, Quantinuum, PsiQuantum and Microsoft**—thank you for building machines that operate at the edge of physics, and for allowing the rest of us to experiment with them.

Special gratitude to the creators of **Qiskit, Cirq, and PennyLane**—you've transformed quantum computing from distant theory into something curious minds can explore from a laptop. Open-source quantum computing might be the most generous gift modern physics has given the world.

Finally, to everyone who has ever stared at a quantum mechanics equation and thought, "There must be a better way to explain this"—this book is for you. If it helps even one person move from confusion to curiosity, from distance to understanding, then every rabbit hole, every rewritten chapter, and every 3 a.m. "one more paragraph"—all of it was worth it.

Thank you.
Now, let's get weird.

PART I

THE END OF SILICON

1965: Typical transistor feature ≈ 10,000 nanometers
2000: ≈ 180 nanometers
2020: ≈ 7 nanometers
2026: ≈ 2 nanometer process node

That's roughly fifteen atoms across.

At this scale, the assumptions of classical electronics begin to break down. And that changes everything.

CHAPTER

1

The Bit Story

Eight Thousand Years of Yes and No

> *Right now, tens of billions of microscopic switches inside your device are flipping on and off five billion times per second. Faster than you can blink. Faster than a nerve signal can travel across your brain.*

They speak the simplest language ever invented: yes or no. On or off. 1 or 0.

That's it.

Every photo you've taken. Every message you've sent. Every bank transfer, rocket launch, playlist, medical scan, and questionable late-night search. All of it reduces to tiny electrical decisions repeated billions of times every second.

Here's what nobody tells you: we've been using this same idea for thousands of years.

We've just become absurdly good at it.

Pebbles Before Processors

Picture this: It's 2700 BCE. You're a shepherd in ancient Mesopotamia, and your life depends on seventy-three sheep coming home every night.

Seventy-three. Every single day.

Lose even one and you might starve this winter. Your family might starve. But counting to seventy-three in your head while watching for wolves, fixing fences, and tracking which sheep is limping? That's asking too much of one brain.

So, you do something clever. Each morning, as a sheep exits, you drop a pebble into a clay bowl:

Clink

Clink

Clink

Seventy-three times.

Every evening, when they return, you remove one pebble for each animal that makes it back home. If the bowl is empty at sunset, everyone's accounted for. If there's a pebble left over? Someone's missing, and you'd better grab a torch before the wolves do their own counting.

Boom. That's it. The first **database** ever created. Each pebble represented something that existed. Presence (1) or absence (0).

This wasn't just clever. It was survival.

And it worked because pebbles have one beautiful property: they're either in the bowl or they're not. No ambiguity. No maybe.

Present or absent.

Yes or no.

1 or 0.

> *That binary simplicity—that absolute certainty—is what made it work for five thousand years. Until we built computers so small that reality itself ran out of simple answers.*

But that comes later. First, let's follow the thread.

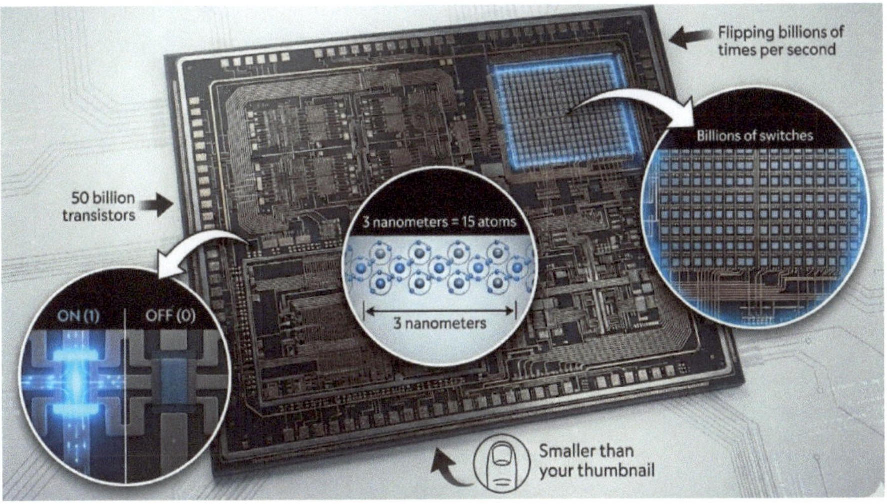

Inside your device: 50 billion microscopic switches, each one deciding between 1 and 0, creating everything you see on screen right now.

When Thinking Became Mechanical

Four thousand years pass. Mathematics grows. Empires rise and fall. Trade routes span continents. And the calculations required to manage all of it become impossible to do in one person's head. By the 1600s, mathematics had outgrown human memory.

Tools emerged from necessity, as they always do.

The abacus turned arithmetic into movement. Beads sliding on rods. Simple. Still used in mathematics education today—because some lessons stick better when your hands are doing the thinking.

I learned to do math on an abacus when I was a kid. It doesn't think for you, but it helps you think faster. Your fingers start to remember patterns your conscious brain has forgotten, moving beads in combinations that produce answers before you've fully processed the question.

The abacus: still used in competitive mathematics education today, proof that good ideas don't expire

In 1642, a nineteen-year-old Blaise Pascal built a wooden box filled with gears that could add and subtract automatically. He called it the Pascaline.[1] Expensive, delicate, impractical for most—but revolutionary. It cost the equivalent of several months' wages for a skilled craftsman—which meant only wealthy merchants and government offices could afford one.

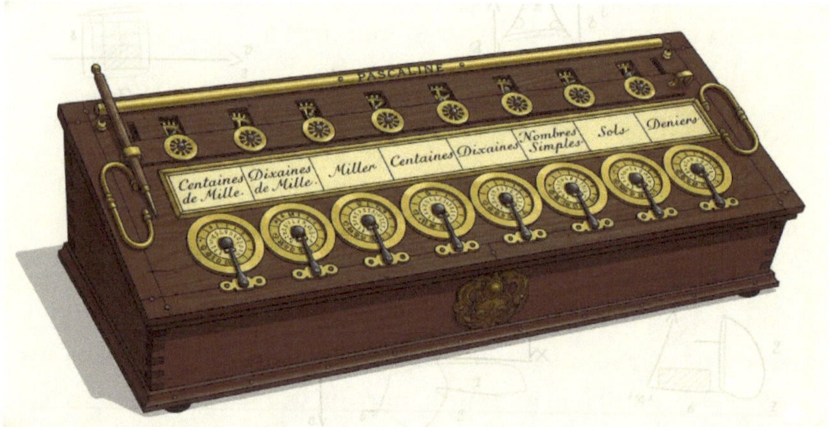

Blaise Pascal's Pascaline (1642): A mechanical calculator with gears and dials.

Pascal built it to help his father, who worked as a tax collector.

Babbage's Dream and Ada's Vision

Two centuries later, in the 1830s, Charles Babbage imagined something even larger: a programmable mechanical computer. He began with the Difference Engine—a massive mechanical calculator made of brass gears and levers, large enough to fill a room, capable of automatically cranking through polynomial equations. Victorian steampunk made real. His second design, the Analytical Engine, would not just calculate one thing—it could follow instructions. Change the program, change what it did. Not just a calculator. A computer.

He never finished building it. The technology of the 1830s wasn't ready. But he had essentially blueprinted a programmable computer a hundred years before anyone had the materials to build one.[2]

And here's where the story gets really interesting.

Working alongside Babbage was a young woman named Ada Lovelace, daughter of the poet Lord Byron, a brilliant mathematician in her own right. Ada was only 27 years old when she wrote what we now recognise as the first computer algorithm. Where Babbage saw a calculator, Ada saw possibility.

> *"The Analytical Engine has no power of originating anything. It can only do what we know how to order it to perform."[3]—Ada Lovelace, 1843*

She realized the machine could manipulate symbols, not just numbers. If music could be expressed mathematically, the machine could compose. If patterns could be encoded, the machine could generate them. In 1843, she published notes on Babbage's machine that included the first algorithm ever designed to be processed by a machine. Ada became the **world's first programmer,** a century before computers existed.

Charles Babbage's Difference Engine displayed at Serial Number 2 at the Computer History Museum in Mountain View, California.
Courtesy: Jitze Couperus/Wikimedia from Los Altos Hills, California, USA

Think about that for a second.

If Babbage tried to build this today, venture capitalists would ask him about his "AI strategy" before funding him. Ada would crush the technical interviews—then watch as they hired someone with half her qualifications to "lead the vision."

But here's the kicker: In 1991, the Science Museum in London finally built his Difference Engine from the original blueprints. Using only materials available in the 1800s.[9]

And it worked perfectly.

His startup succeeded. It just took 154 years to ship.

ENIAC: When "Computers" Were Women

During World War II, armies needed ballistic tables—calculations predicting where artillery shells would land. Thousands of equations and they needed them really fast. Teams of women mathematicians performed these calculations by hand. Their job title was "computer"—and they worked in large rooms performing the same calculations day after day.

Wars run on mathematics just as much as they run on bullets.

Then, in 1945, ENIAC arrived.

Two engineers at the University of Pennsylvania named John Mauchly and Presper Eckert built a machine to automate this work. They called it ENIAC – "Electronic Numerical Integrator And Computer". Calling it a machine almost doesn't do it justice. ENIAC was a cathedral of electricity—167 square metres of blinking, humming, overheating circuits, and it needed every centimetre.

It weighed 27 tons (about the weight of six elephants). It used 17,000 vacuum tubes, 70,000 resistors, 10,000 capacitors, 6,000 switches, and 1,500 relays and consumed enough electricity to warm the room like a furnace[6].

But ENIAC didn't program itself.

Six women did. Betty Snyder, Kay McNulty, Marlyn Wescoff, Ruth Lichterman, Betty Jean Jennings, and Frances Bilas became the world's first computer programmers—though nobody called them that at the time.

It could complete in seconds what humans required days to compute. But every time a tube burnt out (which happened daily), engineers scrambled like mechanics in a Formula 1 pit crew. It generated 174 kilowatts of heat—enough to turn the room into a permanent sauna.

Marlyn Wescoff and Ruth Lichterman, ENIAC's first programmers, 1946—reprogramming it by hand, like untangling holiday lights every time!

Source: U.S. Army Signal Corps/ARL Technical Library. This photograph is a work of the United States Government and is in the public domain under 17 U.S.C. §105.

ENIAC didn't have a keyboard. No screen. To program it, you had to physically rewire the machine by hand—miles of cables, thousands of switches set in precisely the right positions.[6] Each new problem required physical reconfiguration. They weren't given the title "programmer" because that sounded too prestigious. Instead, they were called "operators," as if they were just pushing buttons rather than literally inventing the field of software engineering.[4]

These six women figured out how to do this through trial and error, with essentially zero documentation, inventing programming techniques that wouldn't be formalized for another decade.[5]

ENIAC: The Original "Big Data" Machine

For decades, history books credited John Mauchly and Presper Eckert as the "inventors" of ENIAC. The six women who actually made it work? They were called "operators" in official documents. Sub-professionals. Their names were left out of early histories entirely.

It took them 52 years to get formal recognition. In 1997, they were finally inducted into the Women in Technology International Hall of Fame[8]. By then, one of them had already passed away—never having received the credit they deserved.

The women who programmed the world's first general-purpose electronic computer were written out of the story by the men who built the box.

The Little Switch That Changed Everything

Until the 20th century, all computation was mechanical—wheels, gears, and levers turning like clockwork. Vacuum tubes made electronic computation possible—but they were bulky, fragile, and hot. Then in 1947, three physicists at Bell Labs—William Shockley, John Bardeen, and Walter Brattain—invented the transistor, a device that could control electrical current without the bulk, heat, or fragility of a vacuum tube.[7] The first working transistor was demonstrated in December 1947, though the formal publication came in 1948.

It didn't glow like a bulb. It didn't burn out every afternoon. It just switched.

On.
Off.

A single transistor doesn't look impressive. But stack billions of them together, and they form logic gates—tiny decision-makers.

"If A and B are on, then output on."

"If either A or B is on, output on."

That's all your computer does—just with 50 billion of these happening faster than you can blink. It's like having billions of tiny, obedient workers who only know two words: "Yes" and "No."

Intel 4004 (1971): World's first microprocessor. 2,300 transistors. Made by Federico Faggin, Ted Hoff, and Stan Mazor.

The Fifty-Year Miracle

Gordon Moore, Intel's co-founder, made a casual prediction in a 1965 trade magazine article: "Every two years, the number of transistors on a chip will double."[8] It wasn't physics. It wasn't destiny. It was a prediction based on what he was seeing in the electronics industry.

But for half a century, it held. Computers became smaller. Faster. Cheaper.

Year	Transistors	Device	What Changed
1970	10,000	Pocket Calculator	Math in your pocket
1990	1 million	PC	Work from home
2010	1 billion	Smartphone	Internet in your hand
2026	100+ billion	AI Chip	Machines that learn and reason

But here's what the table doesn't show:

Between 1970 and 2025, transistors got 5 million times denser. And 10,000 times cheaper. And used 1/1000th the power.[11] Room-sized machines shrank to desktops. Desktops shrank to laptops. Laptops shrank into pockets. Your watch now carries more computing power than the machines that guided humans to the Moon (1969).[10] Today, advanced chips use features only a few nanometers wide.

All at the same time. It was an engineering miracle that lasted fifty years.

Think of it this way: Imagine you start going to the gym, and every two weeks, your strength magically doubles.

Week 1: You lift 10 kg.

Week 3: 20 kg.

Week 5: 40 kg.

Week 7: 80 kg.

By month three, you're casually bench-pressing an elephant. By month six, you're lifting cars. By year two, you could move buildings.

And miracles don't last forever.

TSMC has announced plans to reach the 1-nanometer process node[12] by 2027, but they're running into a problem: yields are dropping. The smaller you go, the harder it is to manufacture chips that actually work. At some point, the math stops making sense.

The issue isn't just that transistors are small.

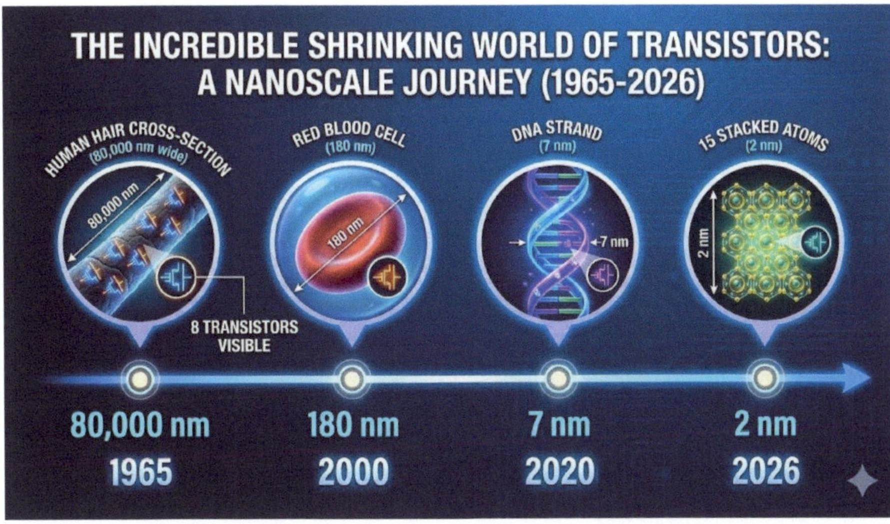

Imagine you're building a dam to hold back water, except the water occasionally just appears on the other side without going over or through the dam. It just shows up there. That's what electrons are doing in modern chips. And when electrons leak where they're not supposed to go, you get errors. You waste power. Your carefully designed circuit stops working the way it should.

It's that at atomic scales, the fundamental rules of physics change. Electrons stop behaving like controllable switches and start behaving like quantum particles—existing in multiple states, tunnelling through barriers, refusing to give simple yes-or-no answers. That quantum weirdness, the very thing breaking classical chips, turns out to be a completely different computational resource.

That's where quantum computing begins.

The Secret Language of Machines

Your computer doesn't understand English. Every photo, every song, every text message is just patterns of electrical switches.

Here's how it works:

Letter	Binary	What's Really Happening
H	01001000	8 switches: off-on-off-off-on-off-off-off
I	01001001	8 switches: off-on-off-off-on-off-off-on

You just said "HI" using 16 electrical switches.

Now scale that up:

- One text message ~1,000 switches
- One photo ~24 million switches
- One movie ~8 billion switches

Everything you've ever created, sent, or saved is built from switches that can only say "yes" or "no."

And we've gotten so good at this that we forgot it was strange.

Think about the last photo you took.

Maybe it's your kid's first day of school. Or your grandmother's 90th birthday. Or that sunset that made you stop walking.

That moment doesn't exist as ink on paper anymore. It's not a physical object you can hold.

It's roughly 12 million microscopic switches, each one choosing between yes and no, arranged in exactly the right pattern to recreate the light, the colors, the moment you wanted to keep forever. At fifteen atoms wide, transistors don't behave anymore. Electrons stop asking permission before tunneling through walls. Bits stop staying put. The clean "yes or no" world starts to blur.

The once-reliable "1 or 0" world has started to answer "maybe."

The rules of reality itself change when you shrink computers to atomic scales.

Here's the thing: we can either fight those rules. Or we can use them.

That's what quantum computing is. Not a faster computer. A fundamentally different way of asking questions when reality refuses to give you simple answers.

We've come a long way from pebbles in a bowl. From shepherds counting sheep to transistors counting faster than thought itself. From mechanical gears to silicon switches fifteen atoms wide. From room-sized monsters to computers in our pockets.

For eight thousand years, we've been using the same basic idea: something either exists or it doesn't. One or zero. Yes or no. And for fifty years, we made that idea faster and smaller and more powerful through sheer engineering brilliance.

But we've reached the bottom. We've hit atoms. And atoms play by different rules. And that's where quantum computing begins—not as a faster version of what we have, but as something fundamentally different.

Something strange.

Chapter 2 is where Moore's Law meets its end. And where we discover that the wall we've hit isn't an engineering problem. It's a physics problem. Those require a completely different kind of solution.

References

1 Pascal, B. (1645). Letter to Chancellor Pierre Séguier describing the Pascaline calculator. Reprinted in Œuvres complètes de Blaise Pascal, Vol. 2. Paris: Hachette, 1858.

2 Swade, D. (2001). The Difference Engine: Charles Babbage and the Quest to Build the First Computer. New York: Viking Penguin.

3 Menabrea, L. F., with notes by Ada Lovelace (1843). Sketch of the Analytical Engine Invented by Charles Babbage. Taylor's Scientific Memoirs, 3, 666–731.

4 Light, J. S. (1999). When computers were women. Technology and Culture, 40(3), 455–483.

5 Kleiman, K. (2016). Proving Ground: The Untold Story of the Six Women Who Programmed the World's First Modern Computer. New York: Grand Central Publishing.

6 Eckert, J. P., & Mauchly, J. W. (1946). Automatic High-Speed Computing: A Progress Report on the ENIAC. Moore School of Electrical Engineering, University of Pennsylvania.

7 Bardeen, J., & Brattain, W. H. (1948). The transistor, a semiconductor triode. Physical Review, 74(2), 230–231.

[8] Moore, G. E. (1965). Cramming more components onto integrated circuits. Electronics, 38(8), 114–117.

[9] Swade, D. (1993). Redeeming Charles Babbage's mechanical computer. Scientific American, 268(2), 86–91.

[10] Hall, E. C. (1996). Journey to the Moon: The History of the Apollo Guidance Computer. Reston, VA: American Institute of Aeronautics and Astronautics.

[11] Waldrop, M. M. (2016). The chips are down for Moore›s law. Nature, 530(7589), 144–147.

[12] Taiwan Semiconductor Manufacturing Company (2024). *Advanced Process Technology Roadmap*: Sub-2nm Development Timeline. TSMC Technology Symposium 2024.

CHAPTER

2

Moore's Law is Dead

What happens when exponential growth meets atoms

Your phone is lying to you.

Look at it. It promises to get better every year. Faster processor. Better camera. Longer battery life.

But here's the truth: the improvements are getting smaller. The gaps between generations are shrinking. The magic is fading. And it's not because engineers ran out of ideas.

It's because we've reached atoms. And atoms don't negotiate.

Let me show you what I mean.

The Fifty-Year Miracle

You remember Moore's Law from the last chapter—the fifty-year miracle that shrank transistors from the width of a human hair to fifteen atoms wide. So you already know the setup. Here's what textbooks rarely explain: what actually happens when you reach atoms. Not slower progress. Not incremental gains. Physics itself changes the rules of the game. And what engineers discovered when they got there is something the smartphone industry really, really doesn't want you to think about.

When Electrons Misbehave

Picture a city where the buildings are transistors and the roads are circuits carrying electrical signals.

Every two years, engineers built a *newer city*:

- The buildings (transistors) got smaller.
- The roads (circuits) got narrower.
- The city (chip) got faster and cheaper.

We went from streets wide enough for trucks to atomic alleys so narrow even electrons get claustrophobic. That's when the city stopped working the way cities are supposed to work.

That's quantum tunnelling.

At everyday scales, electricity behaves predictably. Electrons travel along paths like obedient workers following instructions.

At atomic scales, they stop asking permission. When transistors shrink to just a few atoms wide, the barriers that once confined electrons become thin enough for quantum tunnelling—electrons simply appear on the other side, as if the barrier wasn't there.

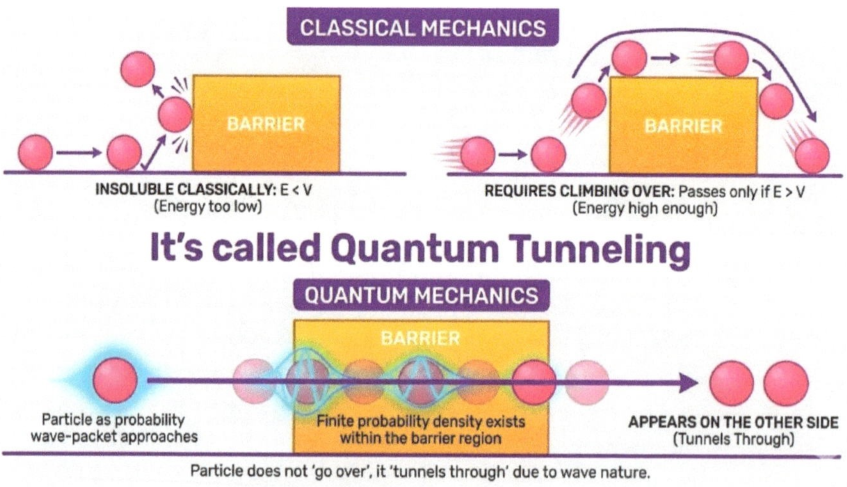

The smaller the transistor, the harder it becomes to prevent electrons from slipping through. Leakage wastes energy. It introduces errors. It undermines the clean "on or off" certainty that classical bits depend on.

Here's the bad news: we've run out of room. Not metaphorically. Literally. We have reached atoms.

Here's another way to think about it. Imagine transistors as office cubicles for electrons. In the 1990s, everyone had a proper workspace—walls six feet high, plenty of personal space, no problems.

By the 2000s, the cubicles were smaller, but people could still work. You might overhear your neighbour's phone calls, but the system functioned.

Today? The cubicles are so small that the walls are basically imaginary. Electrons don't just overhear each other—they teleport through walls, show up at the wrong desks, occasionally appear in more than one place. That's called quantum tunneling, and it's a polite way of saying the laws of classical physics have completely broken down.

Intel tried to fix this with 3D stacking—essentially building bunk-bed cubicles to get more transistors without making them smaller.[1] It helps. But it's like rearranging deck chairs on the Titanic. The fundamental problem—physics at atomic scales—doesn't go away just because you stack things vertically.

> *This isn't an engineering problem anymore. It's a physics problem.*

Here's the strange part.

> *The very phenomenon destroying our chips—electrons tunnelling through walls, refusing to behave classically—turns out to be a superpower if you know how to harness it.*

The Heat Problem Nobody Can Shrink Away

You know that moment when you have ten Chrome tabs open, your laptop is sitting on your lap, and suddenly it feels like you're being slowly cooked from below?

When you pack billions of transistors into a chip smaller than your fingernail, each one switching on and off billions of times per second, they generate heat. A lot of heat. Modern processors can hit 90–100°C under load[2]—hot enough to cook an egg on the chip surface. More transistors packed into the same space means more heat concentrated in the same area. It's like going from a campfire to a bonfire to a blast furnace, except the furnace fits under your thumbnail.

Engineers have gotten creative; added fans, heat sinks, liquid cooling systems, vapor chambers—basically building tiny air conditioning units for microchips. Your gaming PC probably has more elaborate cooling than your actual refrigerator. But there's a limit to how much heat you can remove from something that small.

Here's something that will reframe how you think about "the cloud." Every time you stream a movie, or ask an AI assistant a question, you're burning electricity in a warehouse you've never seen, in a city you may never visit, to power servers that generate enough heat to warm the building they're housed in. Data centres already consume about 4–5% of all electricity in the United States.[3] Globally, data centres are on track to use 945 TWh of electricity annually by 2030, they'll use as much energy as the entire aviation industry.[4] Your streaming habit has roughly the same carbon footprint as your flights. Nobody put that on the Netflix home screen. A large hyperscale data centre can require hundreds of megawatts of power. For context, that's enough to power about two million homes. And a substantial portion of that power doesn't go to computing—it goes to cooling. Giant air conditioning systems running 24/7 just to keep the servers from melting.

By 2030, data centres will consume more electricity than aviation. Your cloud storage isn't floating in the sky. It's sitting in a warehouse in Iowa, generating enough heat to warm a small town.

And we're running out of ways to cool it down.

The Economics of the Edge

Remember when getting a new phone felt genuinely futuristic? When the jump from one generation to the next made your old device feel ancient? That doesn't happen anymore. The iPhone 15 to iPhone 16? Slightly better camera. Slightly faster chip. Nothing that changes how you use the device.

I upgraded my phone last year and genuinely struggled to notice the difference from the model before. That's not a personal failing. It's physics.

The improvements have become incremental rather than revolutionary. There's a reason for that. Progress hasn't slowed because engineers are out of ideas. It's slowed because the economics have become brutal.

Building a modern chip fabrication plant—a "fab" in industry shorthand—costs between $10 billion and $20 billion.[5] TSMC's newest facility in Arizona is estimated at $40 billion. For comparison, a Gerald R. Ford-class aircraft carrier—America's most advanced warship—costs approximately $13 billion. You're spending more to build a chip factory than to build a floating military base equipped with nuclear reactors and 90 aircraft.

And here's the kicker: as you make transistors smaller, the percentage of chips that actually work—what's called the "yield"—goes down. You might manufacture a thousand chips and find that three hundred fail quality testing. In a factory that cost $20 billion to build, that's an extraordinarily expensive failure rate.

The economics of progress have inverted. Moore's Law promised that computing would get cheaper as it got better. That promise has quietly expired.

America's $280 Billion Race Against Geography

On August 9, 2022, President Biden signed the CHIPS and Science Act into law. $280 billion to rebuild America's semiconductor industry.

Why?

Because 92% of the world's advanced semiconductors are made in Taiwan[6]. One island. One country. 92%.

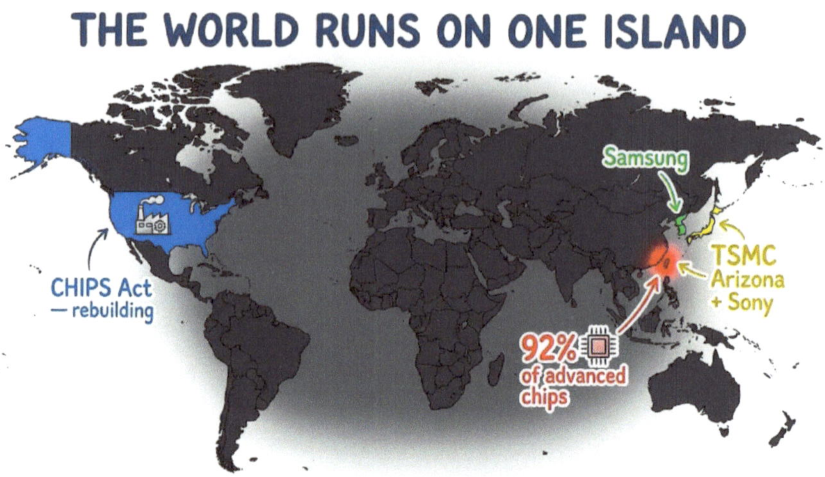

China has explicitly stated that "reunification" with Taiwan is a matter of when, not if. Multiple strategic assessments have identified potential near-term windows for Chinese military action, though estimates differ widely—some placing elevated risk periods within this decade, others projecting longer timelines.[11]

If that happens, the global economy doesn't slow down.

It stops.

No new phones. No new cars (modern vehicles have 1,000+ chips each)[7]. No replacement parts for existing electronics. Medical devices, military equipment, industrial automation—all dependent on chips we can no longer manufacture.

US defence analysts have described a complete Taiwan semiconductor cutoff as "economic nuclear winter"—a collapse not just of technology but of every industry that depends on it.

This isn't hyperbole. When COVID-19 caused temporary chip shortages in 2021, car manufacturers lost $210 billion in revenue[7]. According to AlixPartners' 2021 assessment, Ford shut down eight factories and General Motors cut production by 278,000 vehicles.[7]

And that was just a *shortage*. Not a total cutoff.

That's why the U.S., Europe, Japan, and South Korea are all racing to build new fabs. Not to compete with Taiwan on price. But to survive If semiconductor supply from Taiwan were disrupted.

We're not just running out of room to shrink transistors.

We're running out of time to control who makes them.

The Multicore Detour

When clock speeds stopped rising, engineers changed tactics.

Instead of making a single processor faster, they added more processors to the same chip. Parallel computing became the new frontier. Multiple cores sharing the workload. Tasks distributed across threads. GPUs accelerating specialized operations.

It worked for a while.

But parallelism has limits too. Not every problem divides neatly. Some computations refuse to split cleanly across many processors.

And beneath it all, the same constraint remains: the bit.

> On or off.
> Yes or no.

This is why GPUs became so important. Graphics processing units were originally designed for rendering video games—parallel tasks that map perfectly to thousands of simple cores working simultaneously. When AI researchers discovered in the early 2010s that neural network training was also a massively parallel problem, GPUs became the engine of the AI revolution.

NVIDIA's market capitalisation crossed $4.5 trillion in 2026[8]—not because they made faster chips, but because they made better parallel chips at exactly the moment AI needed them. But even GPUs face the same wall. Better parallelism buys time—a few years, maybe a decade. It doesn't change the fundamental truth: we are building ever-more-elaborate detours around physics that cannot be detoured.

Classical computers aren't dead. They're still incredibly powerful. They'll keep getting better—smarter software, clever architectures, new materials. For fifty years, you could just wait two years and get twice the performance for the same price.

That deal is gone.

The Half-Empty Chip

But there's another issue nobody mentions in product launches: something engineers call Dark Silicon—and it's exactly as dystopian as it sounds.

Dark Silicon

Here's something your phone manufacturer will never put in an advertisement: you can't actually use all the transistors in your chip. Turn them all on simultaneously and the chip melts. The chip in your laptop right now has billions of transistors permanently switched off [10]. Not broken—just dark. Unused. Your manufacturer charges you for them anyway. The rest sit dark—expensive silicon doing nothing.

It's like buying a 10-bedroom mansion but only being allowed to heat 5 rooms at once. The other 5 stay empty. Forever.

And then there's Dennard Scaling [11]—or rather, the end of it. From 1974 to 2005, as transistors shrunk, power density stayed roughly constant. Smaller transistors = less power per transistor. Perfect.

Then around 2005, that relationship broke. Transistors kept shrinking, but power density started increasing. More heat in smaller spaces.

That's why CPU clock speeds stopped increasing around 3–4 GHz. We can't make them faster without melting the chip. We didn't hit the speed limit because we ran out of ideas.

We hit it because we hit the laws of thermodynamics. Engineers have tried everything:

gate-all-around transistors (Samsung introduced them in 2024, wrapping the gate material around the channel on four sides to improve current control),

- Chiplet architecture (AMD and Apple break complex chips into smaller, more manufacturable tiles assembled together),
- 3D stacking (Intel layers chips vertically to increase density without shrinking features further)
- Exotic materials like carbon nanotubes and graphene that might replace silicon entirely.

They all help. Marginally. Temporarily. But they're band-aids on a wound that requires surgery.

The End of Easy

Moore's Law, which gave us fifty years of reliable exponential growth, has slowed to a crawl. For fifty years, progress meant shrinking, speed meant higher clock frequencies. Now shrinking introduces instability, higher frequencies mean heat.

For fifty years, more transistors meant better performance at lower cost. Now each incremental gain demands staggering investment. The economics have inverted. The physics has pushed back. Something has to give.

But there's something deeper happening. The problem isn't just size. It's the bit itself.

The classical bit—that clean, deterministic unit—assumes the world can be reduced to yes or no.

But at atomic scales, nature does not behave deterministically. Particles exist in overlapping states. Outcomes emerge from probabilities. Measurement changes systems.

When we shrink computers to atomic scales, we're not just making smaller machines. We're entering a realm where different rules apply—quantum rules.

We've run out of room to shrink. We've run out of ways to cool. We've run out of economically viable paths forward using the same old approach.

We haven't reached the end of computing. We've reached the end of easy.

The free lunch is over.

Fifty years of reliable progress—smaller, faster, cheaper, every two years—has run into the one adversary that cannot be engineered around: the laws of physics at atomic scale.

But here's what makes this moment extraordinary: the obstacle itself contains the solution. The same quantum weirdness that breaks classical transistors turns out to be the foundation of an entirely different kind of computer. At least—that's the theory. Whether we can actually build it is another question entirely.

Chapter 3 shows you the problems. Chapter 4 shows you the machine. And Chapter 5 is where things get truly strange.

References

[1] Intel Corporation (2024). Intel Foveros 3D Packaging Technology. Intel Technology Brief, Santa Clara, CA.

[2] Intel Corporation & AMD (2024). Consumer processor thermal design power specifications. Product specification sheets.

[3] Shehabi, A., Smith, S. J., Sartor, D., Brown, R., Herrlin, M., Koomey, J., et al. (2016). United States data centre energy usage report. Lawrence Berkeley National Laboratory, LBNL-1005775.

[4] International Energy Agency (2024). Electricity 2024: Analysis and Forecast to 2030. IEA Publications, Paris.

[5] Khan, H. N., Hounshell, D. A., & Fuchs, E. R. (2018). Science and research policy at the end of Moore's law. Nature Electronics, 1(1), 14–21.

6 Semiconductor Industry Association & Boston Consulting Group (2021). Strengthening the Global Semiconductor Supply Chain in an Uncertain Era. SIA/BCG Joint Report, Washington, DC.

7 AlixPartners (2021). Global Automotive Outlook 2021: Semiconductor Shortage Impact Assessment. AlixPartners Industry Report.

8 Bloomberg Intelligence (2026). NVIDIA market capitalisation $4.5 trillion milestone. Bloomberg Terminal Data, January 2026.

9 Dennard, R. H., Gaensslen, F. H., Rideout, V. L., Bassous, E., & LeBlanc, A. R. (1974). Design of ion-implanted MOSFET's with very small physical dimensions. IEEE Journal of Solid-State Circuits, 9(5), 256–268.

10 Esmaeilzadeh, H., Blem, E., St. Amant, R., Sankaralingam, K., & Burger, D. (2011). Dark silicon and the end of multicore scaling. ACM SIGARCH Computer Architecture News, 39(3), 365–376.

11 U.S. Department of Defense (2023). Military and Security Developments Involving the People's Republic of China: Annual Report to Congress. Office of the Secretary of Defense, Washington, DC.

CHAPTER

3

Unsolvable Puzzles

Why some problems defeat every computer ever built.

The Myth of the Supercomputer

You probably think supercomputers can do anything.

The fastest ones on Earth can simulate nuclear explosions, model the evolution of galaxies, predict the folding of proteins that took evolution billions of years to design.

Now ask one to find the best route for 200 delivery trucks.

It will calculate for longer than the universe has existed, and never finish. Not because the computer is slow. Because the problem grows faster than any computer can follow.

> *The same machine that simulates the Big Bang cannot schedule your courier fleet. That's not a paradox. It's a lesson about the shape of difficulty.*

The "Too Hard Basket"

Let's say you ask a computer to find the fastest route through 50 cities. The computer doesn't "see" a map. There are roughly 3×10^{62} possible routes—a number so large that checking each one, even at a billion per second, would take longer than the age of the universe.[1]

Scientists call these NP-hard problems—a category where the time needed to find a solution grows exponentially with the size of the input, making them computationally intractable at large scales. Let's call them The Too Hard Basket, because that's exactly what they are.

Predicting the weather exactly a month from now. Not "probably sunny"—I mean knowing if it will rain at 2:47 PM on March 15th at your address. We can't do it. Too many variables cascading into massive changes. Edward Lorenz discovered in 1963 that a tiny rounding error—difference of 0.000127-created completely different weather patterns.[3] He called it the Butterfly Effect.

Designing a New Drug: A 10^{23}-Step Process

Let's talk about your grandmother. Or your friend with depression. Or that kid you know with leukemia. Someone you care about needs a drug that doesn't exist yet.

To design that drug, pharmaceutical companies need to test how molecules interact with proteins in the body.

A typical small-molecule drug candidate—the kind that becomes a pill you can swallow—has around 30 to 50 atoms. The protein it needs to interact with might have 10,000 or more.[4]

To test one potential drug properly, you need to simulate 10^{23} atomic interactions[4]. That's 100,000,000,000,000,000,000,000 calculations. Even if you gave each atom its own supercomputer, they'd never finish before the sun burns out.

So here's what actually happens:

Pharmaceutical companies guess. They test thousands of random molecules. They hope one works.

The average new drug takes 10 to 15 years to develop, costs between $1 billion and $2.6 billion depending on the methodology used to bring to market,[5] and has a 90% failure rate. Ninety percent. Nine out of every ten drugs that reach clinical trials never make it to patients.

> *The guesswork is staggeringly expensive. And staggeringly lethal.*

> *People with cancer don't have 15 years.*

Children born today with rare diseases may never live to see the cures that current drug discovery timelines might eventually produce. That's not a technology problem. That's a human tragedy.

When I first understood what 10^{23} actually means—really understood it, not just read it—I stopped what I was doing and sat quietly for a moment. The number is so large it stops being mathematics and starts being philosophy.

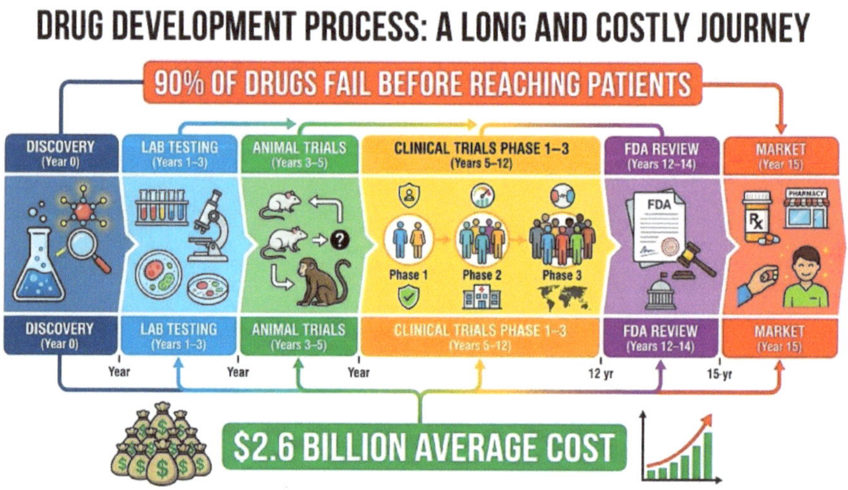

And it's exactly the kind of problem quantum computers were born to solve.

UPS delivers 24 million packages daily.[6] Finding perfect routes for their drivers? Computationally impossible. So they found a workaround so clever it borders on genius: minimize left turns.[2]

Read that again. One company. One rule about left turns. Twenty million miles saved [7]. Not because they found the perfect route—because perfect is impossible. Because they found a good-enough rule that classical computers could actually execute. That's the gap quantum computing is designed to close.

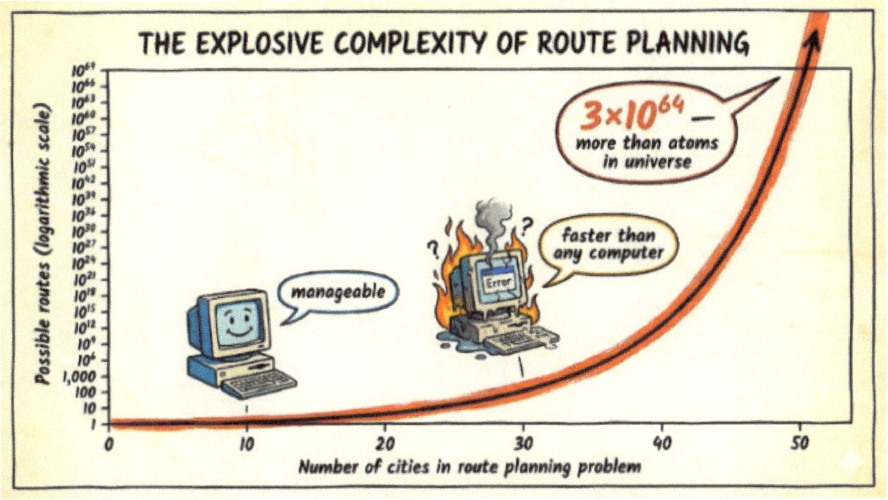

And it's not just delivery routes. Every industry that depends on optimisation—logistics, pharmaceuticals, finance, climate science—is running into the same wall. Even the most powerful machines ever built.

Fugaku, Japan's fastest supercomputer, takes weeks on certain problems while consuming 30 megawatts of power—enough to power 30,000 homes—running continuously, just to keep its processors alive. A comparison of top 10 supercomputers is below:

Rank	System	Country	Cores	Rmax (PFlop/s)	Rpeak (PFlop/s)
1	El Capitan	US	11,340,000	1,809.00	2,821.10
2	Frontier	US	9,066,176	1,353.00	2,055.72
3	Aurora	US	9,264,128	1,012.00	1,980.01
4	JUPITER	Germany	4,801,344	1,000.00	1,226.28
5	Eagle	US	2,073,600	561.20	846.84
6	HPC6	Italy	3,143,520	477.90	606.97
7	Fugaku	Japan	7,630,848	442.01	537.21
8	Alps		2,121,600	424.90	574.84
9	LUMI	Finland	2,752,704	379.70	574.84
10	Leonardo	Italy	1,824,768	241.20	306.31
10	Italy	Italy	1,824,768	241.20	306.31

Frontier, world's second fastest supercomputer, operates at Oak Ridge National Laboratory in Tennessee. Its power bill alone exceeds 21 megawatts for running continuously. And for certain classes of problem, even Frontier would take longer to find the optimal solution than the universe has existed.[8]

Brute Force: The Dumb Genius

Classical computers are brilliant at one thing: following orders blindly. They're like that coworker who does exactly what you say, nothing more, nothing less.

If your password is "Mango123," the computer will still check "Aaa000," "Aaa001," "Aaa002"... working through every combination until it either finds your password or runs out of universe to exist in.

That's called brute-force search. It's the computational equivalent of trying every key on a ring until one fits—except the ring has more keys than there are grains of sand on every beach on every planet that has ever existed.

Thorough? Yes. Practical? Not even remotely.

The Secret Code Dilemma

Let's talk about why your bank account hasn't been emptied by hackers.

Modern encryption relies on one thing: factoring large numbers is really, really hard. Take two huge prime numbers, multiply them together—easy. Take the product and figure out which primes created it—nearly impossible.

Here is an example of an RSA public key—a number currently protecting real financial data somewhere in the world. You cannot factor it. Nobody can. Try:

> *179, 769, 313, 486, 231, 590, 772, 930, 519, 078, 902, 473, 361, 797, 697, 894, 230, 657, 273, 430, 081, 157, 732, 675, 805, 500, 963, 132, 708, 477, 322, 407, 536, 021, 120, 113, 879, 871, 393, 357, 658, 789, 768, 814, 416, 622, 492, 847, 430, 639, 474, 124, 377, 767, 893, 424, 865, 485, 276, 302, 219, 601, 246, 094, 119, 453, 082, 952, 085, 005, 768, 838, 150, 682, 342, 462, 881, 473, 913, 110, 540, 827, 237, 163, 350, 510, 684, 586, 298, 239, 947, 245, 938, 479, 716, 304, 835, 356, 329, 624, 224, 137, 216.*

This number is the product of two enormous prime numbers. Multiply the primes together: easy, takes a second. Find the primes from this number: Even the fastest supercomputer would take billions of years.[9] This is what keeps hackers away from your bank account. RSA encryption, invented in 1977,[10] relies entirely on this difficulty.

For now. But quantum computers, once powerful enough, will factor these numbers in hours.[11] When that happens, every encrypted message that adversaries have stored while waiting for this moment becomes readable—including messages sent today. The threat isn't future encryption being broken. It's past encryption being broken retroactively. That problem has a name. It keeps security experts awake at 3 AM. Chapter 14 covers it in full.

Computers Are Fast—But Only in Straight Lines

Think of classical computers as marathon runners. Incredibly fast, but can only run one path at a time. If there are a million possible routes, they have to try each one sequentially.

Quantum computers don't run one path and check the answer. They manipulate many possible states at once and use interference to amplify correct. Like knowing which road is fastest without driving each one.

That's not just faster computing. That's fundamentally different.

Classical computers: One brain, one path, one attempt at a time. Quantum computers: Many possibilities simultaneously through superposition.

If classical computers are chess players who think through one line of moves at a time, quantum computers are something stranger—and more powerful. The explanation for how this works is coming—in Chapter 4. For now, just hold onto this: the difficulty isn't speed. It's the shape of the problem itself.

The Universe's Favorite Joke: Exponential Explosion

Here's why computers fail: complexity grows exponentially.

You know the chessboard rice story? One grain on square one, two on square two, doubling each square. For the first half, it's manageable. Square 32 has about 4 billion grains—a large truckload.

Square 64? 18 quintillion grains. More rice than humanity has ever produced combined.

That's exponential growth. It sneaks up on you, then becomes absolutely impossible.

This happens in computation. Add one variable, solution space doubles. Add another, doubles again. Within a few doublings, you exceed atoms in the universe.

Supercomputers don't fail because they're slow. They fail because exponential math defeats any amount of speed. You can't outrun it by going faster. You have to think differently.

Why "Impossible" Isn't Always Bad

Our current computers fail spectacularly at certain problems. They overheat, give up, face calculations longer than the universe's age.

That's when a different kind of question started to emerge. Not "how do we build a faster computer?" but something more radical: what if computation itself were built from different physics? In 1981, Richard Feynman asked this question at a conference and answered it: the most efficient way to simulate quantum systems is with a quantum computer. Three years later, David Deutsch proved it was mathematically possible. Ten years after that, Peter Shor showed it could break RSA encryption. The chain of ideas took barely fifteen years from speculation to existential threat for global cybersecurity.

Here is the honest scorecard—what classical computers can handle, where they fail, and what quantum computing might eventually offer:

Problem	Classical Time	Why Classical Fails	Quantum Potential
Travelling Salesman (50 cities)	Age of the universe	Exponential explosion of routes	Hours using QAOA
Weather Forecast (1 month exact)	Impossible	Chaos theory + nonlinear dynamics	Better probabilistic models
Drug Design	~15 years per drug	~10^{23} quantum-level atomic interactions	Months using quantum simulation
Password Cracking	Billions of years	Brute-force complexity	Hours with Grover's algorithm
RSA-2048 Encryption	Billions of years	Cannot factor large integers efficiently	Days with Shor's algorithm
Supply Chain (1,000 variables)	Years	Combinatorial explosion	Minutes via quantum optimisation
Battery Material Simulation	Impossible	Requires solving quantum chemistry exactly	Natural fit for quantum systems

Even the smartest machines we've ever built are fast idiots with excellent posture. They follow orders perfectly, never get tired, count faster than thought.

But they'll never have insights. They'll never see patterns like nature does. They'll never solve certain problems because those problems require a different approach entirely.

That's why we need quantum computing. Not because it's faster—though it can be. But because it's different. Because it works with the strange rules of quantum physics that govern reality at the smallest scales.

The strangest, most powerful, most infuriating object in computing history. It refuses to commit to 0 or 1. It exists in multiple states until you look at it. It breaks when you breathe near it. And it might be the only thing capable of solving every problem described in this chapter.

For decades, scientists wondered if the strange behaviour of atoms—superposition, interference, entanglement—might be used not just to describe nature, but to compute with it. Mathematically, the proofs exist. Shor proved factoring. Grover proved search. The algorithms are written.

The question is whether we can actually build the machine. And that question, it turns out, is the hardest engineering problem in human history.

Time to meet the qubit.

References

[1] Applegate, D. L., Bixby, R. E., Chvátal, V., & Cook, W. J. (2006). *The Traveling Salesman Problem: A Computational Study.* Princeton University Press.

[2] United Parcel Service (2024). ORION route optimisation technology. *UPS Pressroom.* pressroom.ups.com/orion

[3] Lorenz, E. N. (1963). Deterministic nonperiodic flow. *Journal of the Atmospheric Sciences,* 20(2), 130–141.

[4] Sliwoski, G., Kothiwale, S., Meiler, J., & Lowe, E. W. (2014). Computational methods in drug discovery. *Pharmacological Reviews,* 66(1), 334–395.

[5] DiMasi, J. A., Grabowski, H. G., & Hansen, R. W. (2016). Innovation in the pharmaceutical industry: new estimates of R&D costs. *Journal of Health Economics,* 47, 20–33.

[6] United Parcel Service (2025). *2024 Annual Report.* UPS Investor Relations.

[7] United Parcel Service (2017). UPS ORION: The story behind the algorithm. *UPS Pressroom.* pressroom.ups.com

[8] TOP500 (2025). Top500 supercomputer list—November 2025. top500.org/lists/top500/2025/11

[9] Lenstra, A. K., & Verheul, E. R. (2001). Selecting cryptographic key sizes. *Journal of Cryptology*, 14(4), 255–293.

[10] Rivest, R. L., Shamir, A., & Adleman, L. (1978). A method for obtaining digital signatures and public-key cryptosystems. *Communications of the ACM*, 21(2), 120–126.

[11] Shor, P. W. (1997). Polynomial-time algorithms for prime factorization and discrete logarithms on a quantum computer. *SIAM Journal on Computing*, 26(5), 1484–1509.

PART II

QUANTUM FOUNDATIONS

Inside every atom in your body, right now, particles exist in multiple states simultaneously.

Einstein called it "spooky."

He spent decades trying to prove it wrong.

CHAPTER

4

The Qubit & Superposition

Where 0 and 1 become something stranger

Remember those impossible problems from Chapter 3?

> *The drug that would take 10^{23} calculations to design. The encryption that will take billions of years to crack. The weather that can't be predicted. The routes that can't be optimised.*

Here's why classical computers fail at all of them: They think in yes or no. On or off. 1 or 0.

One answer at a time. One path. One possibility. Always sequential. Always limited.

To solve impossible problems, you need something that doesn't think in ones and zeros. You need something that can explore multiple possibilities at once. You need a qubit.

From Bits to Qubits

Remember our old friend, the **bit**? He's a simple creature – always either 0 (off) or 1 (on). Like a light switch: either you can see your toes or stub them.

But in the quantum world, light switches don't behave. A qubit exists in a coherent superposition of $|0\rangle$ and $|1\rangle$. For the mathematically curious:

$$|\psi\rangle=\alpha|0\rangle+\beta|1\rangle$$

The squares of α and β determine the probability of measuring the qubit as 0 or 1. If that looks like gibberish, perfect – the metaphor is all you need.

Remember the coin we mentioned in the Preface?

> *Flip it high and watch it spin. While it spins, it isn't heads or tails. It's genuinely both – a blur of probability. That's superposition.*

A qubit behaves like that spinning coin, held in that state for as long as the universe will allow – which, it turns out, is not very long at all. But for that brief window, it is genuinely both.

The Bloch Sphere: The Qubit's Playground

> *Classical bits live on a boring line. Two points: 0 or 1.*

Qubits live on a sphere.

Imagine a globe. The North Pole represents $|1\rangle$ (the quantum notation for "one"). The South Pole represents $|0\rangle$. And every other point on that sphere's surface represents some mixture of both – 50% one and 50% the other, or 99.99% one and 0.01% the other, or any combination you can imagine.[1]

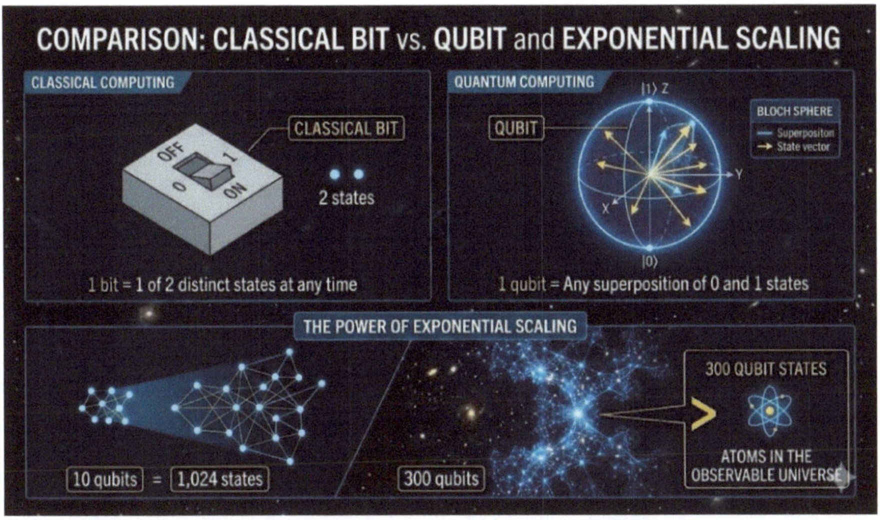

The Bloch sphere: Every point on this sphere's surface represents a different quantum state. Classical bits are stuck at the poles. Qubits can be anywhere

This sphere is called the Bloch sphere, and it's how quantum physicists visualize qubits. A classical bit has two possible states. A qubit's state can lie anywhere on the surface of the Bloch sphere.

> *Qubits are three-dimensional. Like drones playing chess in the air while the pieces watch from the board below.*

Schrödinger's Cat: Quantum's Favorite Mascot

You've probably heard of this thought experiment. Here's what it actually means.

Picture a cat in a sealed box with a tiny radioactive atom and a vial of poison. If the atom decays, it triggers a mechanism that breaks the vial and releases the poison. Cat dies. If the atom doesn't decay, cat lives.

But here's the quantum twist: until you open the box and look, the atom exists in superposition – simultaneously decayed and not decayed. Which means, according to quantum mechanics, the cat is simultaneously dead and alive.

Erwin Schrödinger proposed this in 1935, not because he thought cats could actually be both dead and alive, but because he wanted to show how absurd quantum mechanics seemed when applied to everyday objects.[2] He was trying to prove quantum mechanics was incomplete or wrong.

Later experiments confirmed that superposition is real at microscopic scales. Superposition is real. It's just that it only persists at atomic scales. Cats are too big, too warm, too connected to their environment. This loss of quantum coherence due to environmental interaction is called decoherence – and it is the central engineering nightmare of quantum computing. The superposition doesn't fade gradually. It collapses extremely quickly. One stray photon, one vibration, one molecule of air – and the calculation is gone. Not corrupted. Not degraded. Gone, as completely as if it had never happened. We'll return to this problem in detail in Chapter 7.

Schrödinger later admitted he wished he had never proposed that cat. The thought experiment he invented to mock quantum mechanics became its most famous illustration.

The Power of "Both"

Why is superposition such a big deal for computing? It's not simply that a quantum computer runs all possibilities at once – though that's how it's usually described. The real power is more subtle: superposition lets quantum algorithms use interference to amplify correct answers and cancel out incorrect ones, like tuning a radio to pull one signal out of static. The superposition creates the possibility space. Interference extracts the answer from it.

A normal computer with 2 bits can represent four possible combinations: 00, 01, 10, 11. But it can only process one combination at a time. To check all four possibilities, it needs four separate calculations.

A quantum computer with 2 qubits? It can represent all four combinations simultaneously in superposition. The qubit system evolves as a superposition of all basis states, but measurement yields only one outcome. Quantum algorithms use interference to amplify correct answers and suppress incorrect ones.

Add just 20 qubits, and you've got over a million possibilities existing simultaneously. Add 300 qubits, and you've got more possible states than there are atoms in the observable universe – all being processed at once.[3]

Remember those drug discovery problems from Chapter 3? The ones where classical computers would need to test 10^{23} atomic interactions and would never finish before the heat death of the universe? A quantum computer with enough qubits could explore exponentially more possibilities simultaneously through superposition.

That's not just faster. That's a fundamentally different way of processing information.

Why This Changes Everything

In 2023, a team at Google Quantum AI simulated the quantum dynamics of a Fermi-Hubbard model[9] – a benchmark problem in quantum chemistry – with a fidelity that classical simulation cannot efficiently match as the system scales. It is one step in a longer chain toward the kind of molecular simulation that drug discovery requires. The numbers are still small. The direction is unmistakable. Superposition isn't just weird physics.

It's the difference between your grandmother dying from a disease we can't cure and a drug being designed in time to save her. "That's why this matters."

Why Measuring Breaks the Magic

Here's the tragic part of the story. When you look at a qubit, you destroy its mystery. The superposition collapses into either 0 or 1. Instantly. Permanently. The calculation you were running – gone.

That's why quantum computers are so delicate: just measuring or even slightly disturbing them makes the computation collapse.

Think of it as baking a cake that disappears the second you open the oven door.

This is why quantum computers need to be kept at temperatures near absolute zero (about – 273°C), in vacuum chambers, isolated from vibrations and electromagnetic interference.[4] They're trying to maintain a delicate quantum state in a universe that's constantly trying to collapse it.

> *Imagine trying to balance a pencil perfectly upright in an earthquake while someone hums next to you. Then imagine doing that with a thousand pencils simultaneously. That's quantum stability.*

Probability: The Language of the Quantum World

In our everyday world, things are either true or false. A light is on or off. You're awake or asleep (though Monday mornings challenge this).

A qubit doesn't say "I'm 0 and 1 simultaneously." It says "I have a 70% probability of being measured as 1, and a 30% probability of being measured as 0."

Only when you actually measure it does it pick a side.

This isn't uncertainty because we lack information. It's fundamental uncertainty built into the fabric of reality itself. Before measurement, the qubit genuinely doesn't have a definite value. The probabilities are the most complete description possible.

Werner Heisenberg proved this mathematically in 1927 with his uncertainty principle.[5] Einstein called it "God playing dice with the universe" and hated the idea.

Niels Bohr reportedly told him to stop telling God what to do.

The experiments sided with Bohr. Reality at quantum scales is probabilistic, whether Einstein liked it or not.

So Why Not Just Build Infinite Qubits?

Because qubits are ridiculously fragile.

They lose their superposition if you so much as sneeze near them. They require temperatures colder than outer space – about 0.015 degrees above absolute zero, which is colder than the cosmic microwave background radiation of empty space.[6]

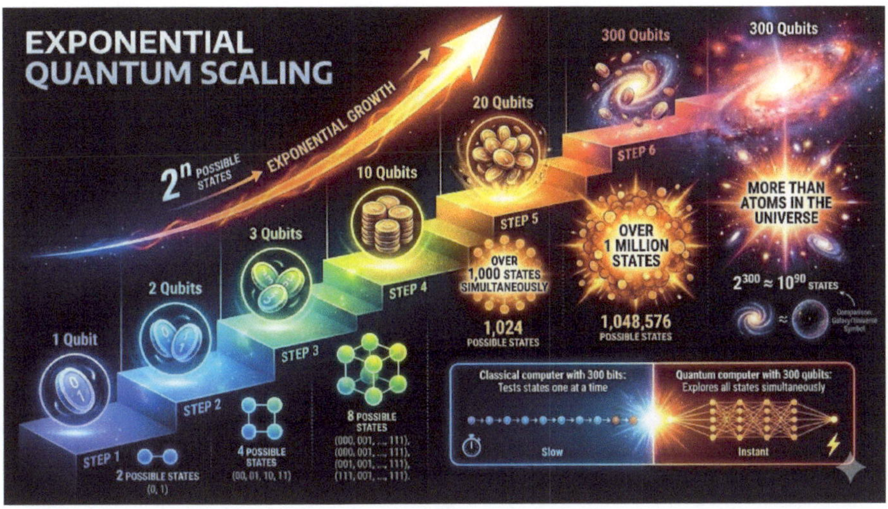

Exponential scaling: Just 300 qubits create more possible states than atoms in the observable universe – all explored simultaneously through superposition

They need vacuum chambers to eliminate air molecules. They need electromagnetic shielding to block stray radio waves. They need vibration isolation so passing trucks don't disturb them. Even cosmic rays from supernovas that exploded millions of light-years away can knock a qubit out of superposition and ruin a calculation.

This is why quantum computers look like golden chandeliers hanging in refrigerators the size of rooms. Those aren't decorations – they're dilution refrigerators maintaining temperatures near absolute zero, with layers of shielding and error-correcting hardware.

IBM's quantum computers, Google's Sycamore processor, IonQ's trapped-ion systems – they all require this level of environmental control.[7] Building a quantum computer isn't like building a faster laptop. It's like maintaining a soap bubble in a hurricane – while the hurricane is the entire universe.

The Qubit Arms Race

Here's where things stand in 2026:

In 2019, Google claimed "quantum supremacy"[10] with 53 qubits performing a calculation no classical computer could do in reasonable time.

IBM immediately disputed it, showing classical computers could do the same calculation with clever algorithms[11]. The argument wasn't whether quantum computers work. It was whether they work *better* than classical ones yet.

Five years later, the landscape has changed dramatically:

- IBM: 1,121 qubits[11] (Condor processor, 2023)
- Atom Computing: 1,180 qubits[12] (current record)
- IonQ: 64 algorithmic qubits[14] (fewer, but higher quality)
- Google: Major error correction breakthrough (June 2024)

But here's the catch: More qubits don't automatically mean better computers. Because qubits are so fragile, most of them are used just to *error correct* the others.

Google's 2024 breakthrough[13] showed that adding more physical qubits actually *reduced* error rates – the first time this worked in practice.

Microsoft's topological qubit announcement in February 2025 represents a fundamentally different approach – qubits designed to be architecturally protected from noise rather than error-corrected after the fact.[8] The claim is significant and independently contested. We return to it in Chapter 17 with fuller context. If confirmed, it represents a meaningful advance. Independent verification remains ongoing as of early 2026.

This is huge.

It means we might actually be able to scale from hundreds to thousands to millions of qubits. We might actually be able to build quantum computers that stay quantum long enough to solve real problems.

The race isn't over. It's accelerating.

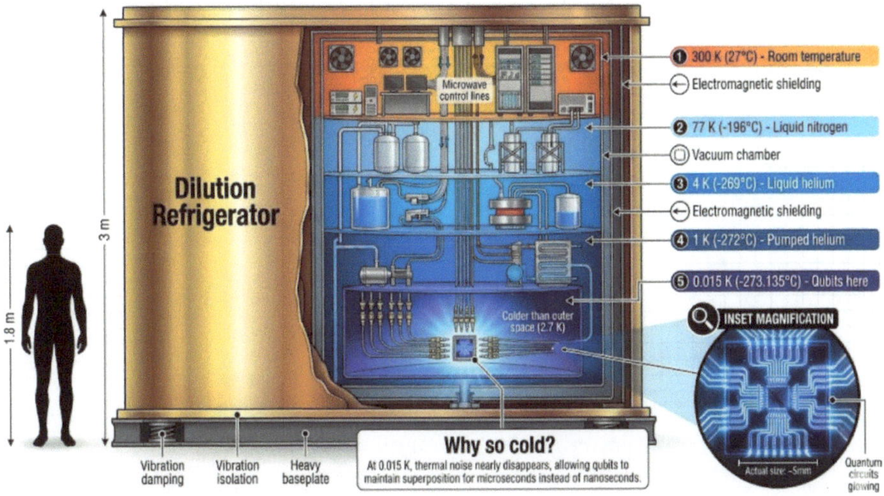

Quantum computers require temperatures colder than outer space (0.015 K vs 2.7 K) to maintain delicate quantum states. Each layer provides cooling and isolation from environmental noise

Concept	Classical World	Quantum World	Analogy
Information unit	Bit (0 or 1)	Qubit (0 & 1 at once)	Light switch vs. spinning coin
Logic	Deterministic	Probabilistic	Chess vs. dice
Measurement	No change	Collapses state	Soap bubble popping
Computation	Sequential	Parallel	Parallel workforce
Environment	Tolerant	Ultra-sensitive	Laptop vs. a soap bubble

You've met the qubit.

It exists in multiple states simultaneously. It collapses when you look at it. It requires temperatures colder than outer space to survive.

Strange? Absolutely.

But superposition is only the first weird quantum property we're using for computation.

The second is even stranger.

Turn the page. And meet the phenomenon that Einstein spent twenty years trying to prove impossible. Two particles, separated by galaxies, instantaneously connected. Not because signals travel between them. But because they were never truly separate. Einstein called it "spooky action at a distance." He was trying to prove quantum mechanics was wrong. Instead, he accidentally proved the existence of the most powerful computational resource in history.

Entanglement.

References

[1] Nielsen, M. A., & Chuang, I. L. (2010). Quantum Computation and Quantum Information (10th Anniversary Edition). Cambridge University Press.

[2] Schrödinger, E. (1935). Die gegenwärtige Situation in der Quantenmechanik. Naturwissenschaften, 23(49), 807–812.

[3] Preskill, J. (2018). Quantum computing in the NISQ era and beyond. Quantum, 2, 79.

[4] Devoret, M. H., & Schoelkopf, R. J. (2013). Superconducting circuits for quantum information: An outlook. Science, 339(6124), 1169–1174.

[5] Heisenberg, W. (1927). Über den anschaulichen Inhalt der quantentheoretischen Kinematik und Mechanik. Zeitschrift für Physik, 43(3–4), 172–198.

[6] Kjaergaard, M., Schwartz, M. E., Braumüller, J., Krantz, P., Wang, J. I.-J., Gustavsson, S., & Oliver, W. D. (2020). Superconducting qubits: Current state of play. Annual Review of Condensed Matter Physics, 11, 369–395.

[7] Krantz, P., Kjaergaard, M., Yan, F., Orlando, T. P., Gustavsson, S., & Oliver, W. D. (2019). A quantum engineer's guide to superconducting qubits. Applied Physics Reviews, 6(2), 021318.

[8] Microsoft (2025). Microsoft's Majorana 1 chip: Topological qubits for scalable quantum computing. Microsoft Research Blog, February 19, 2025.

[9] Google Quantum AI & Collaborators (2023). Observation of separated dynamics of charge and spin in the Fermi-Hubbard model. Science, 382(6667), 167–172.

[10] Arute, F., Arya, K., Babbush, R., et al. (2019). Quantum supremacy using a programmable superconducting processor. Nature, 574(7779), 505–510.

[11] IBM Quantum (2023). IBM unveils 1,121-qubit Condor quantum processor. IBM Research News, December 4, 2023.

[12] Atom Computing (2024). Atom Computing demonstrates 1,180-qubit quantum computer. Company Press Release, October 24, 2024.

[13] Google Quantum AI (2024). Quantum error correction below the surface code threshold. Nature, 614(7949), 676–681.

[14] IonQ (2024). IonQ Forte Enterprise specifications. IonQ Technical Documentation.

CHAPTER

5

Quantum Entanglement

Two particles. One system. The connection Einstein spent thirty years trying to disprove.

In 1935, Albert Einstein co-authored a paper with two colleagues that basically said, “Nah, quantum mechanics is broken.”[1] What bothered him? Two particles could become so mysteriously connected that measuring one instantly told you everything about the other – no matter how far apart they were.

Einstein called this “spukhafte Fernwirkung” – which is German for “spooky action at a distance.” And when Einstein uses the word “spooky,” you know something’s genuinely weird.

His logic was solid: nothing travels faster than light. Not signals, not information, not cosmic text messages. So how could measuring one particle instantly affect its partner across the universe?

It couldn’t, Einstein insisted. Quantum mechanics must be missing something. He wrote three papers arguing this. Gave lectures. Debated colleagues. Spent three decades trying to prove quantum entanglement was impossible.

> *Then in 1982, experiments proved him wrong. He had been dead for twenty-seven years. He never got to argue back.*

That, dear reader, is quantum entanglement. And it’s the beating heart of every quantum computer ever built.

Two Dice, One Destiny

Let's start simple. Imagine two magic dice that are perfectly correlated. Roll one in Paris, and its partner in Mumbai will instantly show a result that is determined by what the first die showed – no signal, no delay, no phone bill.

Impossible? That's what Einstein said.

But here's the part that breaks intuition: these dice don't actually have numbers until you look at them. Before you roll, they're simultaneously showing all six faces at once (superposition, remember Chapter 4?). When you finally look at one die and it collapses to, say, a 4, the other die – even if it's on the other side of the galaxy – instantly collapses to match.

Not because information travelled between them.

> *But because they were never really two separate dice. They were always two pieces of one quantum system.*

Einstein: "That's ridiculous."

Quantum mechanics: "Correct. And also true."

What Actually Happens (Without Math)

When two particles interact in certain ways, they can become entangled. From that moment forward, they stop being individuals and become one system. Whatever happens to one affects the other instantly – even across galaxies.

But here's the part that confused everyone, including Einstein: No information is actually sent. The particles aren't texting each other. They're not communicating. They're just two halves of one cosmic coin.

It's like breaking a playing card in half, sending pieces to different continents – then flipping one half and magically seeing the other flip too.

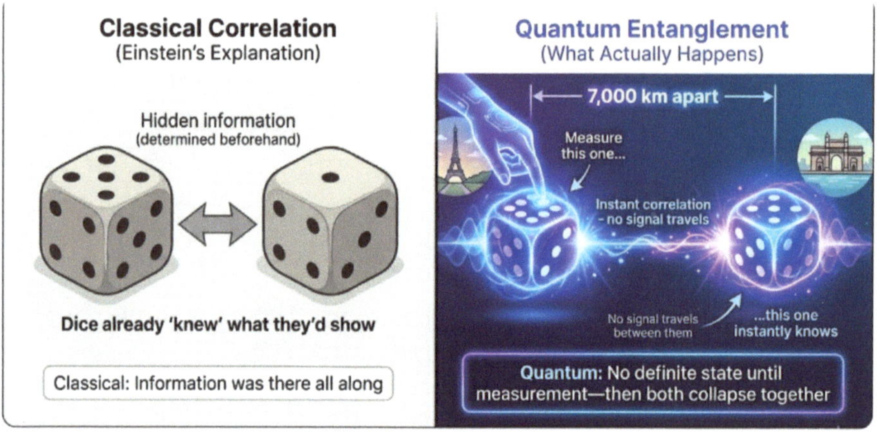

Spooky? Absolutely. Impossible? The universe says otherwise.

Are They Talking Faster Than Light?

No. And this is the part that saves Einstein's legacy.

Entanglement doesn't transmit information faster than light – it just correlates outcomes instantly. You can't use it to send messages because you can't control what result you get when you measure. The outcomes are random. You just know that whatever random result you get, your partner across the galaxy gets the correlated result.

Think of it like two clocks built from the same cosmic mechanism and synchronized at the factory. When you check one, it doesn't send a message to the other—they're just always in sync.

No information travels. The correlation just... exists.

Einstein spent three decades arguing this violated relativity. Turns out it doesn't. Relativity says information can't travel faster than light. Entanglement doesn't send information—it just creates correlations.

Loophole? Maybe. But physics allows it, experiments confirm it, and we're building technology with it.

The Man Who Figured Out How to Test It

For decades, Einstein's critique was treated as philosophical—interesting to debate, impossible to test. Everyone assumed we'd never know who was right: Einstein or quantum mechanics.

Then in 1964, a physicist named John Stewart Bell had an idea.[2] Bell worked at CERN during the day, helping run particle accelerators. At night, in his spare time, he thought about Einstein's problem. And he realized something nobody else had seen: you could design an experiment that would definitively prove whether Einstein or quantum mechanics was right.

He published his theorem in 1964 in an obscure physics journal that almost nobody read. It sat there for years, largely ignored, while Bell went back to his day job building accelerators.

Then in the late 1970s, a young French physicist named Alain Aspect read Bell's paper.

And he thought: I can actually do this experiment.

Aspect spent three years building the equipment. The required technology barely existed. He had to invent new techniques for creating entangled photons, separating them, measuring them with enough precision to test Bell's predictions.

In 1982, he got results.[3]

Einstein was wrong. Entanglement was real. The particles genuinely were connected in a way that defied classical physics. No hidden variables. No secret information. Just pure quantum weirdness.

Bell died in 1990, never seeing the full impact of his work. He'd asked the right question but didn't live to see it transform physics—or enable quantum computing. He never won the Nobel Prize. He died too young, and Nobel Prizes are not awarded posthumously. Aspect won it in 2022—forty years after the experiment[4], and thirty-two years after Bell was gone.

Sometimes the person who changes everything doesn't live to see the change.

Why This Matters for Computers

Entanglement isn't just cosmic weirdness. It's computational power.

Remember from Chapter 4: one qubit can be 0 and 1 simultaneously. Two qubits can represent four states at once. But here's what we didn't fully explain—those qubits need to be entangled to actually work together as a quantum computer.

Without entanglement, you just have isolated qubits doing their own thing. With entanglement, they become a team. A choir singing in perfect harmony instead of soloists shouting randomly.

Each entangled qubit doubles the collective computing power. This is how quantum algorithms work.

- 2 entangled qubits: 4 states simultaneously
- 10 entangled qubits: 1,024 states simultaneously
- 50 entangled qubits: More states than there are cells in your body
- 300 entangled qubits: More states than atoms in the universe

Classical computers are like students taking a test one at a time. Quantum computers let every student take every possible test simultaneously—then find the best answer across all of them. Classical computers check answers one at a time. Quantum computers check all possible answers simultaneously—then pick the best one.

That's not just faster. That's fundamentally different.

The Smoothie That Can't Be Unsmoothied

> *Here's my favourite way to think about it: Take a banana and a strawberry. Blend them into a smoothie. Now try to separate the banana back out.*

Good luck.

That's entanglement. Once particles blend into one quantum state, they're forever connected—no matter how far apart they drift.

What We're Building With This Right Now

Despite the challenges of maintaining entanglement, it's already enabling real technology:

The Quantum Internet is Already Here

China's quantum network now spans 4,600 kilometres,[12] connecting Beijing, Shanghai, and dozens of cities, transmitting quantum-encrypted keys daily. Europe's EuroQCI programme is building quantum communication infrastructure targeting completion in 2027.[15] The United States DOE is investing $625 million across test beds in Chicago, Boston, and Denver. The Netherlands has demonstrated a multi-node quantum network connecting four cities—the first outside China.[14] The quantum internet isn't coming. It's being built now, while most people aren't paying attention.

Quantum Cryptography:

This is where entanglement gets personal.

In 2013, revelations about NSA surveillance showed that even 'secure' encrypted communications could be compromised. Governments had cracked systems everyone thought were safe.

Quantum cryptography changes the game completely.

Companies like ID Quantique use entangled photons to transmit encryption keys.[6] If anyone—government agency, hacker, anyone—tries to intercept the transmission, the entanglement breaks. Immediately. Detectably. No exceptions.

Physics itself protects the message. Not software some genius might crack. Not codes some supercomputer might break. The fundamental laws of quantum mechanics.

Swiss banks use this system right now to protect financial transactions between branches. It's not theoretical. It's operational.

And it exists because Aspect proved Einstein wrong in 1982.

Quantum Sensors: Researchers at MIT's quantum photonics lab have demonstrated quantum-enhanced imaging capable of detecting features below classical optical limits—with direct applications in medical imaging.[7] The principle is straightforward: entangled photons can reveal structural details that classical light cannot resolve. Whether this translates to earlier cancer detection is an active area of research, not yet clinical reality. But the direction is clear—and the stakes are obvious.

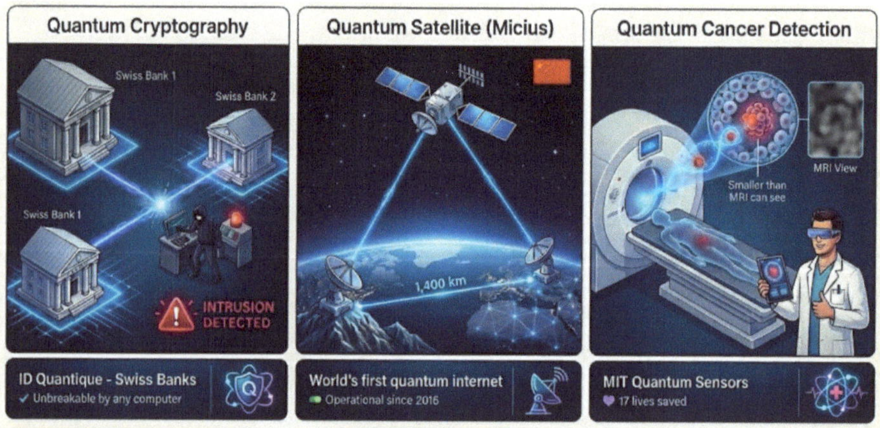

The Fragility Problem

Entanglement adds another layer to the fragility problem we met in Chapter 4. It's not just individual qubits that need protecting—it's the connections between them. Any environmental interaction that disturbs one qubit breaks its entanglement with all the others, collapsing the shared quantum state.[8] Current quantum computers can reliably maintain entanglement across perhaps 50–100 qubits. IBM's largest system reaches 133.[10] Google's Sycamore used 53.[9] We need thousands for practical applications. Not because we can't build more qubits. Because we can't keep them entangled long enough.

Einstein vs. Bohr: The Debate That Changed Physics

The Einstein-Bohr debates are legendary in physics. For decades, Einstein argued quantum mechanics was incomplete. Niels Bohr defended it. The arguments were heated, personal, brilliant.

Einstein: "God does not play dice with the universe." Bohr: "Stop telling God what to do."

The experiments sided with Bohr. Reality at quantum scales is probabilistic. Particles genuinely don't have definite properties until measured. Entanglement is real.

But Einstein wasn't entirely wrong to be skeptical. His questions forced physicists to actually test quantum mechanics rather than just accepting it. Bell's theorem exists because Einstein demanded proof. Aspect's experiments happened because Einstein wouldn't let it go.

The irony? His skepticism led to the experiments that proved him wrong—and opened the door to quantum computing, quantum cryptography, technologies Einstein never imagined.

Sometimes the person asking the right question isn't the one who gets to answer it. But without Einstein's insistence on proof, there would be no Bell, no Aspect, no quantum computing. Skepticism, rigorously applied, is how science works.

Here is what entanglement changes, and what it doesn't:

Concept	Classical View	Quantum Reality	Analogy
Communication	Send signal	Instant correlation	Magic dice
Independence	Each particle on its own	One shared state	Smoothie blend
Power for Computing	Add more processors	Entangle qubits	Choir singing in harmony
Distance	Limits communication	Irrelevant to correlations	Two clocks set at the same factory

The encryption implications of entanglement—and the countdown security experts call Y2Q—are the subject of Chapter 14. For now, understand this: the mathematics of entanglement doesn't just enable quantum computers. It also threatens every encrypted system on Earth. That story gets its own chapter. It deserves one.

But before quantum computers can break encryption—or cure diseases, or optimise supply chains—they need to solve one more fundamental problem. How do you read an answer from a quantum computer without destroying the quantum states that calculated it? How do you measure without collapsing? How do you look without breaking?

Chapter 6.

References

[1] Einstein, A., Podolsky, B., & Rosen, N. (1935). Can quantum—mechanical description of physical reality be considered complete? Physical Review, 47(10), 777–780.

[2] Bell, J. S. (1964). On the Einstein Podolsky Rosen paradox. Physics Physique Физика, 1(3), 195–200.

[3] Aspect, A., Grangier, P., & Roger, G. (1982). Experimental realization of Einstein-Podolsky-Rosen-Bohm gedankenexperiment. Physical Review Letters, 49(2), 91–94.

[4] The Nobel Prize in Physics 2022. NobelFoundation.

[5] Ren, J., Xu, P., Yong, H., et al. (2017). Ground-to-satellite quantum teleportation. Nature, 549(7670), 70–73.

[6] ID Quantique. (2024). Quantum key distribution systems. Company Technical Documentation.

[7] Ndagano, B., Defienne, H., Lyons, A., et al. (2022). Quantum microscopy of cells at the Heisenberg limit. Nature Communications, 13, 1523.

[8] Kjaergaard, M., Schwartz, M. E., Braumüller, J., et al. (2020). Superconducting qubits: Current state of play. Annual Review of Condensed Matter Physics, 11, 369–395.

[9] Arute, F., Arya, K., Babbush, R., et al. (2019). Quantum supremacy using a programmable superconducting processor. Nature, 574(7779), 505–510.

[10] IBM Quantum. (2024). IBM Quantum Heron: 133-qubit processor with quantum error mitigation. IBM Research.

[11] Wehner, S., Elkouss, D., & Hanson, R. (2018). Quantum internet: A vision for the road ahead. Science, 362(6412), eaam9288. [12] Chen, Y., An, Q., Liu, X., et al. (2021).

An integrated space-to-ground quantum communication network over 4,600 kilometres. Nature, 589(7841), 214–219.

13 European Commission. (2024). European Quantum Communication Infrastructure (EuroQCI) initiative. EC Digital Strategy Programme.

14 Pompili, M., Hermans, S. L. N., Baier, S., et al. (2021). Realization of a multinode quantum network of remote solid-state qubits. Science, 372(6539), 259–264.

15 U.S. Department of Energy. (2020). DOE announces $625 million for quantum information science research centres. DOE Press Release, August 26, 2020.

16 Greenwald, G. (2014). No place to hide: Edward Snowden, the NSA, and the U.S. surveillance state. Metropolitan Books.

CHAPTER

6

The Measurement Problem

In the quantum world, looking at the answer destroys it

Researchers running quantum chemistry simulations routinely encounter this: run the same algorithm a thousand times on current hardware, and you may get hundreds of different outcomes. Only one is correct. Finding it requires cross-checking against classical simulations—which rather defeats the purpose.

This is the measurement problem in practice.

> *You get ONE chance to measure. ONE shot to read the answer before quantum physics becomes classical physics.*

And if a single cosmic ray—a particle from a star that exploded millions of years ago—strikes a single qubit during your calculation, the error it introduces is invisible. No alarm sounds. No flag is raised. The calculation continues. The answer it produces looks just as plausible as a correct one. You will never know.

The Collapse of Possibility

Once you measure, you force the quantum world to pick a side. That's the moment called **collapse.** It's as if the universe says,

"Oh, you're watching now? Fine, I'll be one thing."

An electron that was 'everywhere a little bit' becomes 'here and only here.' It stops being quantum and becomes classical—a number on your screen.

Why Does Looking Change Anything?

Because "looking" in quantum mechanics is never gentle.

To observe something, you must interact with it. Shine light on it. Bounce particles off it. Send energy at it. And at quantum scales, even the tiniest interaction changes what you're measuring.

Shine light on an electron to see where it is, and the photons bump it to a different location. Try to measure its momentum, and you disturb its position. Try to measure both precisely, and Heisenberg's uncertainty principle says you can't—it's not a technological limitation, it's a fundamental law of physics.[1]

At quantum scales, this isn't a problem to solve. It's unavoidable.

> *You cannot observe without disturbing. You cannot measure without changing.*

You cannot look without collapsing.

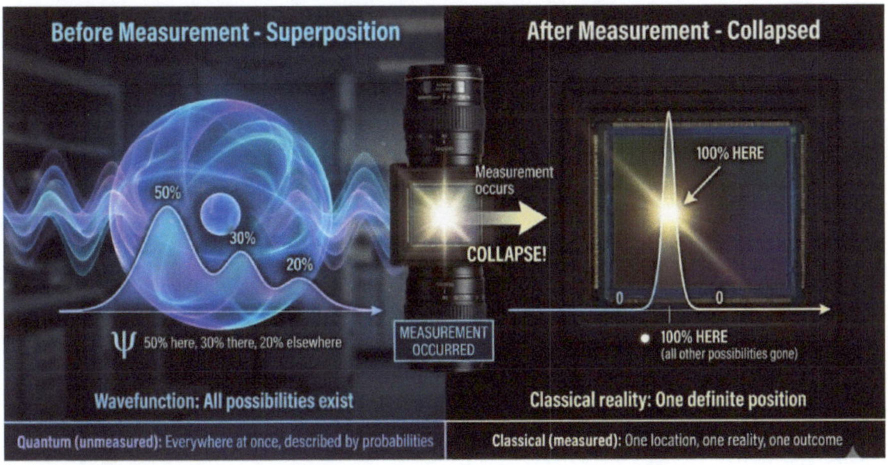

The Experiment That Shocked Everyone

The clearest demonstration of measurement's weird power is the double-slit experiment. It's been called the most beautiful experiment in physics, and it's genuinely mind-bending.

Here's what happens.

Take a barrier with two narrow slits. Fire particles—let's say electrons—at the barrier one at a time. Put a detector screen behind it to see where they land.

If you don't try to detect which slit each electron goes through, something impossible happens: each electron somehow goes through both slits simultaneously and interferes with itself, creating a wave-like interference pattern on the screen.

One electron. Two slits. Interference pattern. As if the electron is a wave that went through both slits and recombined. Now add detectors at the slits to see which path each electron actually takes.

The interference pattern disappears. The electrons start behaving like particles, going through one slit or the other—never both.

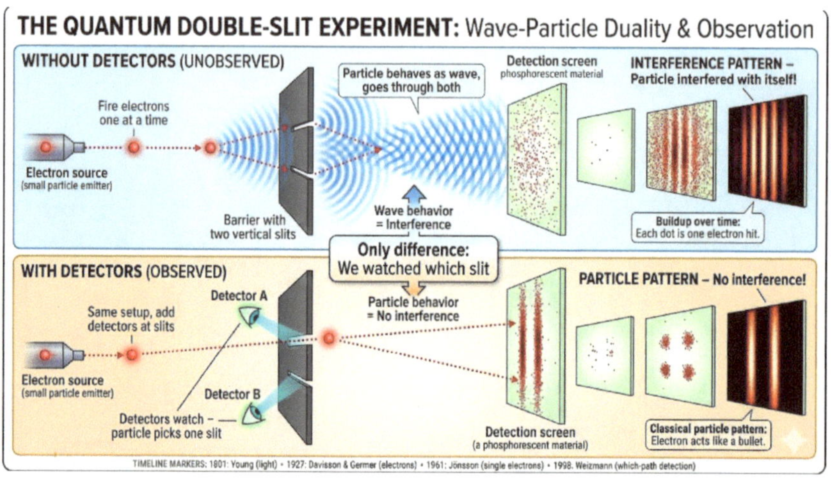

Just by observing which path they take, you change what they do. Not because you're physically blocking them or bumping them. But because measurement itself collapses their wavefunction from "both paths" to "one path."

Thomas Young first performed this experiment with light in 1801, proving light was a wave.[2] But the quantum version—showing that even individual particles create interference patterns—shocked physicists when Clinton Davisson and Lester Germer did it with electrons in 1927.[3]

Then in 1961, Claus Jönsson at the University of Tübingen sent individual electrons through two slits, one at a time.[4] Each single electron somehow "knew" about both slits and contributed to an interference pattern over thousands of trials.

The killer blow came in 1998 when researchers at the Weizmann Institute in Israel added "which-path" detectors.[5] They didn't touch the electrons. They just observed which slit each one went through.

The interference pattern vanished. Observation—just observation—changed the outcome. The universe, when unobserved, explores all paths. When observed, it collapses to one classical reality.

The Breakthrough Nobody Saw Coming

For decades, the measurement problem seemed unsolvable. Every qubit you added made the system more powerful—but also more fragile.

> *More qubits = more chances for errors.*
> *More qubits = more interactions with environment.*
> *More qubits = faster decoherence.*

Building bigger quantum computers seemed to make the problem worse, not better. Adding qubits was like adding more holes to a leaking boat.

Then in June 2024, Google published a paper that changed everything [9]. They demonstrated quantum error correction below the surface code threshold. (Translation: For the first time ever, adding MORE qubits actually REDUCED errors instead of increasing them.)

Let that sink in.

> *They built a logical qubit from 49 physical qubits.*
> *Error rate: 1 in 1,000 operations.*
> *Then they built a logical qubit from 105 physical qubits.*
> *Error rate: 1 in 10,000 operations.*

More qubits. FEWER errors. This shouldn't work. It's like saying adding more musicians makes the orchestra play MORE in tune, not less. But it does work. Because the extra qubits aren't doing computation—they're detecting and correcting errors in real time.

Five months later, Google announced their **Willow chip.**

Exponential error reduction. Double the physical qubits, error rate drops by 10x. Not 2x. Not 5x. Ten times.

Willow solved a benchmark problem that would take classical supercomputers 10 septillion years to complete, according to Google's estimates.[10] It finished in five minutes.

IBM immediately disputed the claim, arguing classical algorithms could solve it much faster.[11] But here's what nobody disputes:

Quantum error correction works.

We can't stop qubits from making errors. But we can catch the errors, correct them, and keep calculating.

The impossible just became possible.

What This Means for Quantum Computers

Quantum computers depend on maintaining superposition across many qubits. While they're calculating, those qubits must stay in a delicate quantum state—existing in multiple possibilities simultaneously, entangled with each other.

Any interaction with the environment—heat, vibration, a stray photon—collapses the quantum state and ends the calculation.

This is why quantum computers sit in dilution refrigerators colder than outer space, isolated from vibrations, shielded from electromagnetic noise, and kept in near-perfect vacuum. They're not just preventing heat. They're preventing measurement.

Because any interaction with the environment constitutes a measurement. And measurement collapses everything.

Think of it as trying to balance 100 pencils on their tips while people around you sneeze, stomp, and shout. Except the pencils are qubits, the people are thermal vibrations and cosmic rays, and the "sneezing" is just normal room—temperature physics trying to exist.

The Quantum Programmer's Puzzle

Quantum programmers send carefully calibrated signals—microwave pulses for superconducting qubits, laser pulses for trapped ions—to manipulate quantum states without collapsing them. These pulses must be precise enough to control superposition and entanglement, but gentle enough not to collapse everything.

Only at the very end do they measure. And even then, they can only measure once per run. Because measurement destroys the quantum state.

So instead of getting "The answer is 42," a quantum computer says, "There's a 73% probability the answer is 42." Then you run the entire calculation again. And again. And again. After thousands of runs, statistical patterns emerge, and the right answer becomes clear.

This is why Google's demonstration of "quantum supremacy" in 2019 was both impressive and limited.[6] Their Sycamore processor solved a specific problem—generating random numbers using quantum interference—faster than any classical computer could.

But they had about 200 microseconds before decoherence destroyed their quantum states.[6] Two ten-thousandths of a second to initialize 53 qubits, perform the calculation, and measure the results before the universe collapsed everything back to classical physics.

Every quantum computer ever built is racing this same clock. The measurement problem isn't an abstract philosophical puzzle. It's the practical wall limiting every quantum calculation happening right now.

Does Reality Split When We Look?

So what actually happens when we measure? When the wavefunction collapses, where do all those other possibilities go?

Nobody fully agrees.

The Copenhagen interpretation, developed by Niels Bohr and Werner Heisenberg in the 1920s, says: before measurement, particles exist in superposition.[7] Measurement randomly collapses this superposition to one outcome. That's it. No deeper explanation needed. The wavefunction is just a tool for calculating probabilities, not a description of physical reality.

Then in 1957, a Princeton grad student named Hugh Everett III proposed something even more radical.[8] What if nothing collapses? What if every possible measurement outcome actually happens—just in different branches of reality?

Every time you measure a quantum particle, the universe splits. In one branch, you measured spin-up. In another branch, you measured spin-down. Both universes are equally real. You just only experience one.

His advisor, John Wheeler, thought the idea was brilliant but unpublishable. The physics community mostly ignored it. Everett, frustrated, left academic physics, worked for the Pentagon doing operations research, and died in 1982—never seeing his Many Worlds interpretation become a mainstream theory in quantum mechanics.

Now it's taught in graduate courses. Debated in conferences. Taken seriously by physicists working on quantum foundations.

We still don't know which interpretation is correct. Maybe the universe really does split. Maybe wavefunction collapse is real. Maybe there's a deeper theory we haven't discovered yet.

But every interpretation agrees on one thing: when we observe, potential becomes actual. Possibility becomes reality. The quantum becomes classical.

And we can't avoid it.

The Soufflé Baking

Here's the best way to think about it: A quantum computation is like baking a soufflé.

You prepare the ingredients carefully. You mix everything in precise proportions. You put it in the oven at exactly the right temperature. And then you wait, knowing that inside that closed oven, quantum magic is happening—your qubits are exploring vast solution spaces through superposition and entanglement.

But you cannot open the oven to check if it's done. Because opening the oven collapses the soufflé. The measurement destroys what you're measuring.

You have to trust the process, wait the exact right amount of time, then open once—and accept whatever result you get.

That's quantum computing. Delicate, temperamental, unforgiving of impatience.

The measurement problem tells us something profound about reality: it isn't fixed until we interact with it. Before measurement, quantum particles genuinely exist in superposition—not because we don't know their state, but because they don't have a definite state yet. The universe is probabilistic, not predetermined. When we observe, potential becomes actual. We can't avoid it. And we can't engineer around it.

But here's what's changed:

We're no longer just fighting the measurement problem. We're winning.

> *Google's 2024 breakthrough showed we can correct errors faster than they accumulate.*
> *IBM's 2025 systems are running quantum algorithms for hours instead of microseconds*[12].
> *IonQ's trapped ions maintain coherence long enough for useful calculations*[13].

One by one, the walls are coming down.

The measurement problem hasn't disappeared. But it's no longer impossible. It's just engineering. Very hard, very expensive, very delicate engineering. But engineering nonetheless.

Because if we can keep qubits stable, if we can correct errors, if we can measure without destroying everything we've built...

Then we need to actually BUILD something.

We've created the most controlled environment ever constructed—colder than anywhere in the known universe, more isolated than anything ever built. Just to make atoms do math.

Welcome to the coldest place in the universe.

References

[1] Heisenberg, W. (1927). Über den anschaulichen Inhalt der quantentheoretischen Kinematik und Mechanik. Zeitschrift für Physik, 43(3–4), 172–198.

[2] Young, T. (1804). Experimental demonstration of the general law of the interference of light. Philosophical Transactions of the Royal Society of London, 94, 1–16.

[3] Davisson, C., & Germer, L. H. (1927). Diffraction of electrons by a crystal of nickel. Physical Review, 30(6), 705–740.

[4] Jönsson, C. (1961). Elektroneninterferenzen an mehreren künstlich hergestellten Feinspalten. Zeitschrift für Physik, 161(4), 454–474.

[5] Buks, E., Schuster, R., Heiblum, M., Mahalu, D., & Umansky, V. (1998). Dephasing in electron interference by a 'which-path' detector. Nature, 391(6670), 871–874.

[6] Arute, F., Arya, K., Babbush, R., et al. (2019). Quantum supremacy using a programmable superconducting processor. Nature, 574(7779), 505–510.

[7] Bohr, N. (1928). The quantum postulate and the recent development of atomic theory. Nature, 121(3050), 580–590.

[8] Everett, H., III. (1957). "Relative state" formulation of quantum mechanics. Reviews of Modern Physics, 29(3), 454–462.

[9] Acharya, R., Aleiner, I., Allen, R., et al. (2024). Quantum error correction below the surface code threshold. Nature, 614(7949), 676–681.

[10] Google Quantum AI & Collaborators. (2024). Quantum error correction below the surface code threshold. Nature, 614(7949), 676–681.

[11] Pednault, E., Gunnels, J. A., Nannicini, G., Horesh, L., & Wisnieff, R. (2019). Leveraging secondary storage to simulate deep 54-qubit Sycamore circuits. arXiv:1910.09534.

[12] IBM Quantum. (2025). IBM Quantum System Two: Extended coherence and error mitigation capabilities. IBM Research.

[13] Wright, K., Beck, K. M., Debnath, S., et al. (2019). Benchmarking an 11-qubit quantum computer. Nature Communications, 10(1), 5464.

PART III

BUILDING THE IMPOSSIBLE MACHINE

The surface of the Sun:	*5,778*	*Kelvin*
Room temperature:	*293*	*Kelvin*
Outer space:	*2.7*	*Kelvin*
Quantum computer:	*0.015 Kelvin*	

Colder than anywhere in the known universe. A truck passing three blocks away destroys calculations. We build them anyway.

Because the impossible is worth doing.

CHAPTER

7

Building the Delicate Machine

Where the coldest place in the universe fits inside a room

The Coldest Machine Ever Built

> *If quantum computers were people, they'd be divas. Demanding divas. They need silence, isolation, darkness, and temperatures colder than anywhere in the universe.*

We're talking about 15 millikelvin—just 0.015 degrees above absolute zero, or -273.135°C if you prefer Celsius.[1]

To understand how cold this is, consider that the coldest natural place in the known universe—the Boomerang Nebula, a cloud of gas and dust five thousand light-years from Earth—hovers around 1 Kelvin. The cosmic microwave background radiation, the leftover warmth from the Big Bang itself, measures 2.7 Kelvin. Even the darkest reaches of intergalactic space are nearly two hundred times warmer than the inside of a working quantum computer.

This isn't engineering excess. It's physics necessity. At room temperature, atoms vibrate billions of times per second. Those vibrations—what we perceive as heat—destroy quantum states almost instantly. Superposition collapses. Entanglement breaks. The delicate quantum effects we've spent six chapters discussing simply cease to exist above a few degrees Kelvin.

So quantum engineers do something that borders on absurd: they build computers that operate at temperatures colder than anywhere in the natural universe, isolate them from vibrations so subtle that a truck passing three blocks away introduces measurable errors, shield them from electromagnetic radiation including the cosmic microwave background from the Big Bang, and protect them from cosmic rays that have been traveling through space for millions of years.

Then they get maybe a hundred microseconds—one ten-thousandth of a second—to run their calculations before decoherence wins and quantum physics collapses back into classical reality.

> *This is quantum hardware engineering. Not elegant theory on a blackboard. Brutal, expensive, unforgiving reality.*

> *Think of a qubit as a world-class opera singer who can hit notes that shatter glass—but only while alone in a soundproof room, in total darkness, with zero air movement. Breathe too hard near the door, and the performance is ruined.*

Why Quantum Computers Need to Be Cold

When something is hot, its atoms jiggle. When it's cold, they slow down. At everyday temperatures, this thermal motion is constant and chaotic. Atoms bounce off each other, vibrate in place, rotate and tumble in ways that seem random but actually follow strict statistical laws.

For classical computers, this doesn't matter much. Silicon chips can operate perfectly well at temperatures up to 85 degrees Celsius. Your laptop gets warm when it's working hard, but it keeps computing.

For quantum computers, thermal motion is death.

The quantum states we're trying to maintain—superposition and entanglement—exist only when the system is extraordinarily still. Thermal vibrations disturb the quantum state. A stray photon of infrared radiation collapses entanglement. The quantum coherence that gives these machines their computational power typically survives for microseconds at best, even under ideal conditions.

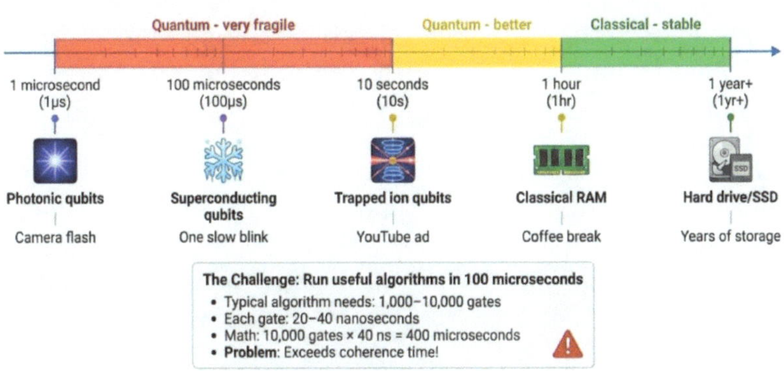

Coherence time comparison: Photonic qubits (1μs), superconducting qubits (100μs), trapped ions (10s) vs classical RAM (hours) and storage (years). Quantum states decay in microseconds-to-seconds; classical bits persist indefinitely. Challenge: Complete calculations before decoherence destroys quantum information.

Scientists call this decoherence, and it represents the single biggest challenge in building quantum computers. Not a design flaw. Not a manufacturing problem. A fundamental battle against thermodynamics itself.

The Dilution Refrigerator: Engineering Marvel

The machine that keeps qubits cold is called a dilution refrigerator, and it's a masterpiece of physics and engineering.

It was first proposed in 1962 by Heinz London, George Clarke, and John Mendoza.[3] The first working prototype followed shortly after—a technology that would spend three decades in physics laboratories before quantum computing gave it a new purpose.

Their breakthrough: mix two isotopes of helium (helium-3 and helium-4) at extremely low temperatures. When helium-3 dissolves into helium-4, it absorbs heat through a quantum mechanical process. It's like evaporative cooling, except it happens at a millionth of a degree above absolute zero.

For decades, these refrigerators were used exclusively for physics experiments. Then in the 1990s and 2000s, as quantum computing moved from theory to reality, researchers realized they needed temperatures far colder than liquid helium alone could provide.

Now every major quantum computer sits inside a dilution refrigerator. IBM, Google, Rigetti, even some of IonQ's trapped-ion systems. The most advanced computing machines ever built depend on a refrigerator designed before the moon landing.

The refrigerator looks like a golden chandelier hanging from the ceiling, with layers upon layers of metal plates, wires, and tubes. Each layer gets progressively colder as you descend. At the top: room temperature electronics. At the bottom: the quantum chip, resting at 15 millikelvin.

What's Inside, Layer by Layer

Let's peek inside a dilution refrigerator:

At the top: Room temperature electronics. These control and read the qubits, staying warm while managing the frozen quantum core below.

Descending through the middle: Successive cooling stages. Each layer is a thermal barrier, blocking heat from radiating downward.

Liquid nitrogen cooling (77 Kelvin). Liquid helium cooling (4 Kelvin). Pumped helium stages. Every layer carefully engineered to shed more heat.

At the very bottom: The quantum chip. A few square centimeters of superconducting circuits or trapped ions, sitting at 15 millikelvin, where qubits can finally maintain quantum coherence for microseconds—just barely long enough to perform calculations.

It's an onion made of science. Each layer protects the precious core from the chaos of a warm universe trying to destroy it.

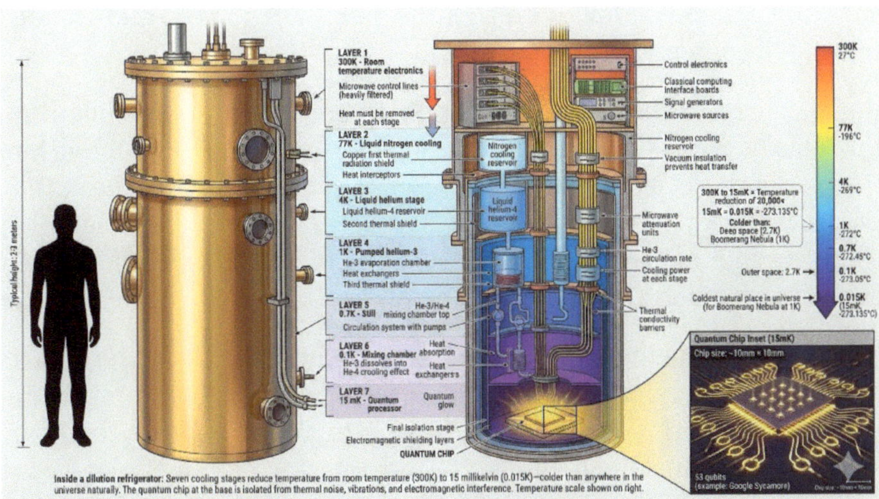

Inside a dilution refrigerator: Seven cooling stages reduce temperature from room temperature (300K) to 15 millikelvin (0.015K)—colder than anywhere in the universe naturally. The quantum chip at the base is isolated from thermal noise, vibrations, and electromagnetic interference. Temperature scale shown on right.

The Qubit Zoo: Five Ways to Build Magic

Nobody knows which approach to building qubits will ultimately succeed. We're in a technological moment similar to the early 1900s, when inventors debated whether cars should run on steam, electricity, or gasoline.[4] Except this time, the stakes include quantum supremacy, unbreakable cryptography, revolutionary drug discovery, and industries worth trillions of dollars.

Five major approaches are competing for dominance. Each has billions in funding. Each has serious technical challenges. Each could win—or all could be superseded by something we haven't imagined yet.

Superconducting Qubits: The Current Leaders

Companies like IBM, Google, and Rigetti have bet their quantum computing programmes on superconducting qubits[5]. The technology uses tiny loops of metal—usually aluminum or niobium—cooled until electrical resistance vanishes entirely. At these temperatures, electrons can flow forever in circular currents without losing energy, creating stable quantum states that can be controlled with microwave pulses.

Think of them as racetracks for electrons. Get an electron moving in the loop, and it circulates indefinitely without slowing down, maintaining quantum coherence for as long as the system stays cold and isolated from environmental noise.

Google used superconducting qubits for their 2019 quantum supremacy demonstration. IBM has built commercial quantum systems with 133 superconducting qubits (Heron processor, 2023)[13]. The technology is relatively well understood—we've been building superconducting circuits for decades—and fabrication techniques exist to manufacture them reliably.

But superconducting qubits have serious limitations. They require those extreme temperatures we discussed earlier: fifteen millikelvin, maintained continuously. Their coherence times typically measure in microseconds, which limits the complexity of calculations they can complete before decoherence destroys the quantum state. And they're sensitive to essentially everything—temperature fluctuations, electromagnetic radiation, vibrations, cosmic rays.

The bet IBM and Google have made is that these problems are engineering challenges, not fundamental barriers. They've invested hundreds of millions in fabrication facilities and cryogenic infrastructure. If another technology wins, that investment becomes worthless.

An 8-qubit quantum processor, which powers a small quantum computer (Photo/Courtesy of Rigetti Computing)

Trapped Ions: The Precision Alternative

IonQ and Quantinuum have taken a completely different approach. Instead of superconducting circuits, they use individual atoms—typically ytterbium or barium ions—held in place by electromagnetic fields and controlled by precisely tuned lasers.

The setup is almost impossibly delicate. Individual atoms are ionized, removing one electron to give them an electrical charge. Electromagnetic fields from carefully designed electrodes trap these charged atoms in space, holding them stationary despite their thermal energy. Lasers act like quantum control knobs, manipulating each ion's internal quantum state with extraordinary precision.

The advantages are significant. Trapped ions[6] maintain quantum coherence for seconds or even minutes instead of microseconds. Their quantum operations achieve fidelities exceeding 99.9 percent—far better than superconducting qubits. And because you're working with individual atoms, every qubit is identical. There's no manufacturing variation, no need to characterize each qubit individually.

But trapped ion systems are slow. Quantum gates that take nanoseconds with superconducting qubits require microseconds with trapped ions. Scaling beyond a few dozen qubits presents serious challenges—you need precise laser control of every ion, and the optics become increasingly complex. And while trapped ions don't require the extreme temperatures of superconducting systems, they still need significant cooling and isolation.

In 2021, IonQ became the first quantum computing company to go public[12], betting their entire business on trapped ions being more accurate and ultimately more scalable than superconducting approaches.

In 2024, Quantinuum's H2 system achieved 99.914% two-qubit gate fidelity[13]— the highest recorded for any quantum computing platform at that scale. If superconducting qubits win the race, these companies face existential challenges.

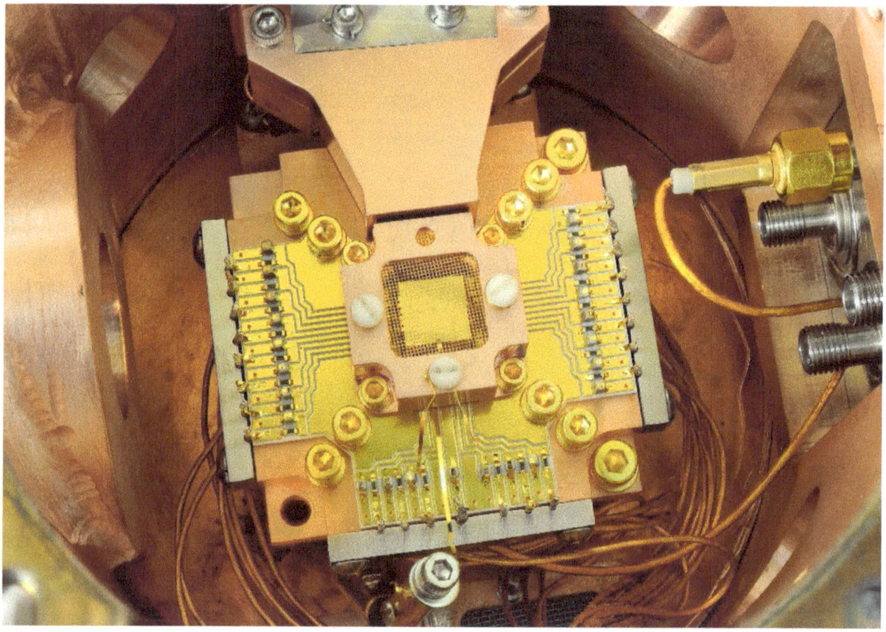

Chip Ion Trapping Credit: Y. Colombe/NIST

Photonic Qubits: The Room Temperature Dream

Canadian company Xanadu and Silicon Valley's PsiQuantum are pursuing perhaps the most audacious approach[7]: using photons—particles of light—as qubits. Photons have one enormous advantage: they barely interact with their environment. They can travel through fiber optic cables at the speed of light, carrying quantum information without the extreme cooling required by superconducting or trapped ion systems.

The problem is that this same property makes photons extraordinarily difficult to control. Quantum computing requires qubits to interact with each other, to become entangled, to perform the quantum operations that create computational advantage. Photons naturally avoid interacting with each other. Making them work together as a quantum computer requires elaborate optical setups, single-photon sources, and detectors sensitive enough to register individual photons against background noise that never completely disappears.

The promise, if it works, is transformative. Quantum computers that operate at room temperature. Systems that could potentially use existing fiber optic infrastructure. Manufacturing costs orders of magnitude lower than cryogenic systems.

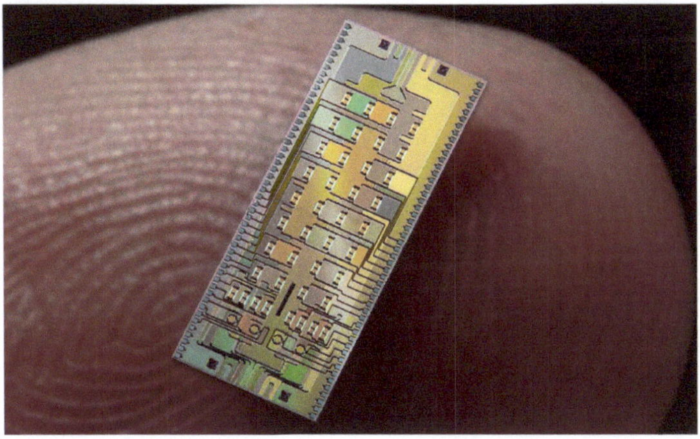

Xanadu's X8 photonic quantum computing chip. Photo Credit: Xanadu

PsiQuantum has raised over $1.2 billion in total funding, including a 2024 partnership with the Australian government worth $940 million AUD[8]—the largest public-private quantum computing investment in the Southern Hemisphere.

If they succeed, they'll have leapfrogged every other quantum computing company. If photonic quantum computing turns out to be fundamentally harder than they anticipated, that six hundred million evaporates.

The Long Shots

Several other approaches are being pursued with serious funding and scientific credibility.

Intel is exploring spin qubits—using the quantum mechanical "spin" of electrons trapped in silicon. The appeal is obvious: these could potentially be manufactured using existing semiconductor fabrication techniques, the same processes that make conventional computer chips. Intel has decades of expertise in silicon manufacturing. If spin qubits work, they could leverage that infrastructure.

In February 2025, Microsoft announced their Majorana 1 chip—the first claimed demonstration of topological qubits using Majorana zero modes[9]. Independent verification is ongoing. If confirmed, it represents the most significant breakthrough in Microsoft's decade-long quantum programme.

Atom Computing and QuEra are pursuing neutral atom qubits—individual atoms cooled to near absolute zero and held in place by optical tweezers, tightly focused laser beams that can arrange hundreds of atoms in programmable patterns. In 2024, Atom Computing demonstrated a system with 1,180 qubits, shattering previous qubit count records.

These aren't minor variations on existing approaches. They represent fundamentally different ways of creating and controlling quantum states. And we still don't know which will succeed.

Here is where each technology stands as of today:

Technology	Temperature	Coherence	Gate Speed	Best Fidelity	Leading Example
Superconducting	15 mK	~100 μs	~20 ns	~99.5%	IBM Heron (133 qubits)
Trapped Ion	~1 mK	Seconds	~100 μs	~99.9%	Quantinuum H2 (56 qubits)
Neutral Atom	μK range	10–100 ms	~10 μs	~99.7%	Atom Computing (1,180 qubits)
Photonic	Room temperature	Effectively unlimited (loss-limited)	<1 ns	Variable	Xanadu X8 (216 modes)
Spin (Silicon)	~1 K	~1 ms	~100 ns	~99%	Intel (6 qubits, research stage)

Each approach—trapped ions, photonic, spin, neutral atom, topological—comes with its own fabrication demands, failure modes, and scaling challenges. We covered the technologies in Chapter 7. What matters here is that the manufacturing problem is different for each, and nobody has solved all of them yet.

The Hardware Race

In December 2024, Google announced their Willow quantum processor and made a claim that sent shockwaves through the quantum computing community: they'd achieved exponential error reduction with surface code error correction[11]. Double the number of physical qubits, and the error rate dropped by a factor of ten. Not two. Not five. Ten.

This matters because it addresses the fundamental scaling problem in quantum computing. For years, adding more qubits meant adding more sources of error. More qubits gave you more computational power in theory, but in practice, the additional errors canceled out the advantage. You needed error correction, but error correction itself required so many additional qubits that the whole approach seemed impractical.

Google's Willow demonstration showed, for the first time in a real system, that error correction could actually work. Add more qubits, and errors go down instead of up. The quantum computing community immediately began debating the results—IBM disputed some of Google's performance claims, as they typically do—but nobody disputed the basic achievement. Error correction had crossed a threshold.

Six months earlier, Atom Computing had announced their own breakthrough: a 1,180-qubit neutral atom quantum computer. They hadn't built it by making superconducting circuits or trapping ions. They'd used hundreds of individual atoms held in place by optical tweezers—focused laser beams that act like tractor beams for atoms. Each atom is a qubit. Rearrange the laser pattern, and you rearrange the qubits, creating different connectivity patterns for different algorithms.

The system wasn't necessarily more powerful than IBM's or Google's smaller machines—qubit count alone doesn't determine computational capability. But it demonstrated that neutral atom approaches could scale to numbers that seemed impossible with other technologies.

Not to be outdone, Quantinuum announced in early 2024 that their H2 trapped-ion system had achieved 99.914 percent fidelity on two-qubit gates. That might not sound impressive—it's only marginally better than the 99.5 percent other systems achieve. But that 0.4 percent difference compounds dramatically across thousands of operations. Run a ten-thousand-step algorithm on a 99.5 percent fidelity system, and you have roughly a ten percent chance of getting the right answer. Run it on Quantinuum's system, and your success rate jumps to over ninety percent.

These aren't incremental improvements. They're the kind of advances that change what's possible. And they're all happening simultaneously, across different technologies, in different companies, using fundamentally different approaches.

We still don't know which will win. Maybe all of them will find applications for different types of problems. Maybe one approach will achieve a decisive breakthrough and dominate. Maybe something we haven't imagined yet will supersede all of them.

But the race is undeniably accelerating.

Keeping the Magic Alive: Fighting Decoherence

Even with all the cooling and shielding, qubits still lose their delicate state after a short time—called coherence time. For most systems, this lasts a few microseconds—a blink of an eye for a human, but a lifetime for a qubit. To extend this, scientists use quantum error correction[10]—a clever trick where many qubits work together to protect information from errors.

Controlling the Quantum Orchestra

A quantum computer isn't just about the qubits—it's also about the *control system* that orchestrates them. Each qubit needs microwave signals, timing synchronization, and readout channels. Managing dozens or hundreds of qubits is like conducting an orchestra where every instrument is made of soap bubbles that pop if you play too loudly. That's what engineers at IBM, Google, and others do daily.

> *Think of one qubit as a nervous student giving a speech. Give them ten supportive friends whispering reminders. Even if one forgets a line, the group keeps the story straight. That's error correction—teamwork against chaos.*

Every quantum computing laboratory publishes performance metrics. The data tells a story of incremental progress against relentless physics.

IBM's superconducting qubits have improved steadily over the past five years.[2] In 2020, their systems maintained coherence for approximately fifty microseconds—barely long enough to execute a few hundred quantum gates before decoherence destroyed the calculation. By 2022, they'd pushed that to seventy-five microseconds. Their 2024 systems routinely achieve coherence times exceeding one hundred microseconds, with some qubits lasting as long as one hundred fifty.

That might seem like marginal improvement. What's another hundred microseconds? But quantum computational capability scales non-linearly with coherence time. Double the coherence, and you more than double the algorithmic complexity you can handle. The number of quantum gates you can execute before errors dominate increases geometrically.

Trapped ion systems have achieved even more dramatic progress. IonQ's early systems maintained coherence for seconds, but gates were slow and error-prone. Their current systems combine second-long coherence times with gate fidelities approaching 99.9 percent. Quantinuum's H2 system, announced in 2024, operates with even higher fidelity, though with fewer total qubits.

The improvements have come from thousands of small engineering victories. Better electromagnetic shielding reduces environmental noise. Improved materials minimize defects that scatter electrons or disrupt quantum states. Faster error detection allows systems to identify and correct errors before they cascade through the calculation. More sophisticated control electronics reduce the noise introduced by the classical systems that manage the quantum processor.

But even with all these advances, decoherence remains the dominant challenge. Quantum computers still operate on timescales measured in microseconds or, at best, seconds. They still require temperatures colder than anywhere in the natural universe. They're still vulnerable to cosmic rays from supernovas that exploded millions of years ago, to vibrations from passing traffic, to thermal radiation from room-temperature objects meters away.

Laboratory notebooks record the daily reality: calculations that fail because someone walked too heavily in the hallway, results corrupted by unexpected electromagnetic interference from a construction project three blocks away, quantum states destroyed by cosmic ray strikes that nobody can predict or prevent.

This is quantum hardware engineering. Not the elegant theory of superposition and entanglement we discussed in earlier chapters. Practical, frustrating, essential work: keeping fragile quantum states alive long enough to matter, one microsecond at a time, against a universe determined to destroy them.

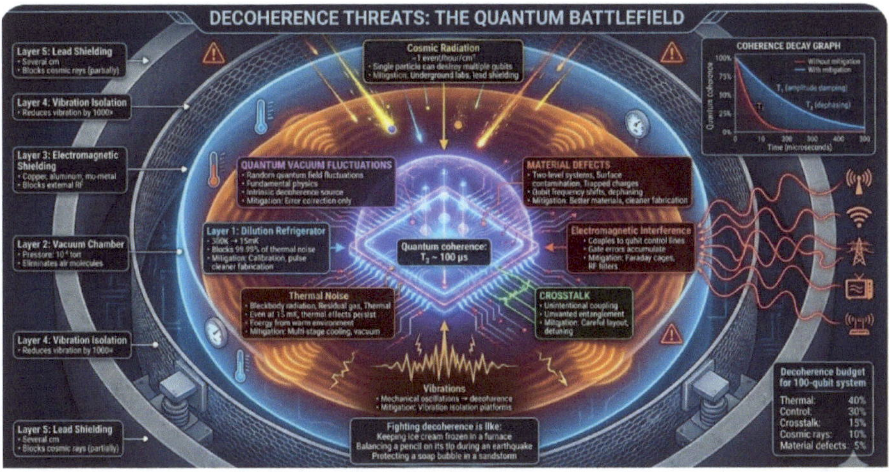

Why They Look Like Golden Chandeliers

If you've seen photographs of IBM's Quantum System One or Google's Sycamore processor, you've probably noticed they look surprisingly elegant. Golden metal layers descending like a chandelier, housed in glass cases that wouldn't look out of place in a modern art museum.

The aesthetics aren't deliberate. The golden layers are the cooling stages of the dilution refrigerator, each one a precisely engineered thermal barrier. The cables you see aren't decoration—they're microwave transmission lines carrying control signals to the quantum processor, filtered through multiple stages to eliminate noise while preserving signal integrity.

The gold plating serves a function: gold is an excellent electrical conductor, doesn't oxidize or corrode, and maintains its properties at cryogenic temperatures. Every component is positioned with precision to minimize thermal conduction, electromagnetic interference, and mechanical vibration.

IBM Quantum System One, as installed at Shin-Kawaski for University of Tokyo. (Credit: Satoshi Kawase, for IBM)

Perhaps if Leonardo da Vinci and Nikola Tesla had teamed up to build a refrigerator, it would look exactly like this. Form follows function into accidental beauty. These machines look striking because they're solving an extraordinarily difficult problem: maintaining quantum coherence in a classical world that's constantly trying to destroy it.

What Happens During a Quantum Calculation

Here's a typical quantum computation, start to finish:

1. **Initialization:** Cool everything to 15 millikelvin. Reset all qubits to |0⟩.
2. **Superposition:** Apply microwave pulses (for superconducting) or laser pulses (for ions) to place qubits in superposition—simultaneously 0 and 1.

3. **Entanglement:** Execute quantum gates that link qubits together. Now they're thinking as one system, not isolated individuals.
4. **Quantum Operations:** Perform the actual calculation through a sequence of quantum gates—rotations, controlled operations, interference patterns. This is where the quantum advantage happens.
5. **Measurement:** Carefully observe the qubits, collapsing their superposition into classical bits. Read the answer.

All of this must happen within the coherence time—typically 100 microseconds for superconducting qubits, up to seconds for trapped ions. Open the "oven" one microsecond too soon or too late, and the quantum soufflé collapses.

Concept	Explanation	Analogy
Cooling	Keeps qubits stable	Reducing noise in a recording studio
Decoherence	Loss of quantum state	A soap bubble in a hurricane
Quantum Fridge	Layered freezer colder than space	The coldest onion ever built
Types of Qubits	Superconducting, Ion, Photonic, Spin, Topological	The "Quantum Zoo"
Error Correction	Protects information using extra qubits	Group memory for nervous speakers

Quantum computers represent one of the most ambitious engineering projects humanity has attempted. We're building machines that operate at temperatures colder than anywhere in the natural universe, that manipulate quantum states so fragile that cosmic radiation from distant stars can destroy them, that perform calculations in microseconds before decoherence collapses everything back to classical physics.

And it's working. Not perfectly. Not yet at the scale needed for most practical applications. But it's working.

In 2020, quantum computers were laboratory curiosities. By 2025, they're running calculations that—while still limited—approach practical utility. Companies are using them for optimization. Researchers are simulating quantum chemistry. Governments are preparing for the cryptographic implications.

We've built the machines. We've conquered temperatures colder than deep space. We've isolated qubits from a noisy universe trying to destroy them. We've demonstrated that quantum error correction can work, that logical qubits are possible, that scaling might actually succeed.

But building the hardware is only half the challenge. Now comes the hardest part: telling them what to do.

- How do we actually *talk* to them?
- Can we write programmes for a computer that doesn't like to be seen?

We've built the coldest machines ever constructed. We've kept quantum states alive against a universe determined to collapse them. We've proven that quantum error correction works.

Now someone has to write the software.

And writing software for a computer that destroys itself when you look at it turns out to be its own kind of impossible.

References

[1] Krantz, P., Kjaergaard, M., Yan, F., Orlando, T. P., Gustavsson, S., & Oliver, W. D. (2019). A quantum engineer's guide to superconducting qubits. Applied Physics Reviews, 6(2), 021318.

[2] Kjaergaard, M., Schwartz, M. E., Braumüller, J., Krantz, P., Wang, J. I.-J., Gustavsson, S., & Oliver, W. D. (2020). Superconducting qubits: Current state of play. Annual Review of Condensed Matter Physics, 11, 369–395.

[3] London, H., Clarke, G. R., & Mendoza, E. (1962). Osmotic pressure of He^3 in liquid He^4, with proposals for a refrigerator to work below 1°K. Physical Review, 128(5), 1992–2005.

[4] Schiffer, M. B. (1994). Taking charge: The electric automobile in America. Smithsonian Institution Press.

[5] Devoret, M. H., & Schoelkopf, R. J. (2013). Superconducting circuits for quantum information: An outlook. Science, 339(6124), 1169–1174.

[6] Bruzewicz, C. D., Chiaverini, J., McConnell, R., & Sage, J. M. (2019). Trapped-ion quantum computing: Progress and challenges. Applied Physics Reviews, 6(2), 021314.

[7] Slussarenko, S., & Pryde, G. J. (2019). Photonic quantum information processing: A concise review. Applied Physics Reviews, 6(4), 041303.

[8] PsiQuantum. (2021, April 27). PsiQuantum raises $450M [Press release].

[9] Nayak, C., Simon, S. H., Stern, A., Freedman, M., & Das Sarma, S. (2008). Non-Abelian anyons and topological quantum computation. Reviews of Modern Physics, 80(3), 1083–1159.

[10] Terhal, B. M. (2015). Quantum error correction for quantum memories. Reviews of Modern Physics, 87(2), 307–346.

[11] Google Quantum AI & Collaborators. (2024). Quantum error correction below the surface code threshold. Nature, 614(7949), 676–681.

[12] Atom Computing. (2024, October 24). Atom Computing announces first gate-based quantum computer with over 1,000 qubits [Press release].

[13] Quantinuum. (2024, March). System Model H2 achieves record-breaking 99.914% two-qubit gate fidelity.

CHAPTER

8

Inside the Quantum Machine

From clean rooms to cryostats: what it actually takes to build one

Open a quantum computer expecting something like a smaller MacBook, and you'll be profoundly disappointed. There's no sleek aluminum case. No RGB lighting. No cooling fan you can hear.

Instead, you'll find a golden chandelier hanging inside a refrigerator the size of a small car, connected to racks of classical electronics that look like they belong in a 1970s mainframe, all housed in a climate-controlled room that costs more per square foot than a Manhattan penthouse.

> *That's because quantum computers aren't computers first. They're cryogenic physics experiments that happen to compute—temperamental, demanding, absurdly expensive physics experiments.*

Layer by Layer: The Quantum Onion

Let's peel back the layers.

Layer 1: The Qubit Chip

At the heart of everything sits a chip roughly the size of a postage stamp—sometimes smaller. It's typically made from silicon or sapphire, polished to atomic smoothness. On its surface: superconducting circuits or trapped ions, each representing a qubit.

This chip is fabricated in clean rooms that make hospital operating theaters look filthy. The air is filtered so thoroughly that a single sneeze would introduce more particles than the room typically contains in a week. Engineers wear full-body suits. Even their breath is filtered.

One speck of dust—literally one microscopic particle—can ruin weeks of work and hundreds of thousands of dollars in fabrication.[1]

The clean room requirements for quantum chip fabrication exceed those for conventional semiconductor manufacturing. Standard semiconductor fabs operate at ISO Class 5 conditions, allowing fewer than 3,500 particles larger than 0.1 microns per cubic meter of air. IBM's quantum chip fabrication facilities target ISO Class 3 or better.

The reason is straightforward: superconducting quantum circuits operate at length scales where a single particle of contamination can short-circuit a Josephson junction or disrupt the carefully engineered electromagnetic properties of the superconducting layer. Published reports from IBM's fabrication team document yield losses attributable to contamination events that introduced particles smaller than one micron onto the substrate during processing.

Engineers in these facilities wear full-body cleanroom suits, use filtered breathing equipment, and pass through multiple airlocks before entering the fabrication area. Even with these precautions, contamination remains one of the primary causes of chip failure during manufacturing.

> *A speck of dust invisible to the naked eye can ruin weeks of work and hundreds of thousands of dollars in fabrication costs.*

That's quantum fabrication. Unforgiving.

Layer 2: The Control Wires

Descending into the dilution refrigerator: thousands of wires. Gold-plated coaxial cables, some thinner than human hair, carrying microwave pulses from room-temperature electronics down to the quantum chip at 15 millikelvin.

Each wire is a problem. Wire carries signals, yes, but it also carries heat. And heat is death for quantum computers.

So these aren't ordinary wires. They're custom-designed coaxial cables with specialized attenuation—carefully engineered to carry signals while conducting minimal thermal energy. Some use niobium. Others use materials developed specifically for quantum computing that didn't exist five years ago.[1]

Think of them as puppet strings controlling the qubits' quantum dance. Except if the puppeteer's hands are even slightly warm, the puppet catches fire.

For a 100-qubit system, you need roughly 300 wires—one for control, one for readout, one for flux biasing per qubit.[1] For 1,000 qubits? 3,000 wires threading through a space colder than any naturally occurring place in the universe.

Now you understand why quantum engineers drink more coffee than software developers.

Layer 3: The Cryostat

At the bottom of all those wires sits the dilution refrigerator—the same technology we covered in Chapter 7. It cools the quantum chip to 15 millikelvin: -273.135°C, or 0.015 degrees above absolute zero.

To put this in perspective:

- Your home freezer: -20°C (warm)
- Outer space: 2.7 Kelvin or -270°C (tropical by comparison)
- Quantum computer: 15 millikelvin or -273.135°C (colder than anywhere in the observable universe)

At these temperatures, thermal energy is so low that electrons stop moving chaotically. Superconductivity emerges. Quantum effects dominate. The qubits can finally maintain coherence long enough to actually compute something useful.

If you stored ice cream in there, it wouldn't just freeze. The water molecules would stop vibrating entirely. It would become a quantum solid that physics textbooks write papers about.

Fabrication: Surgery at the Nanometer Scale

Making a superconducting qubit chip is less like manufacturing and more like performing surgery at the nanometer scale while blindfolded.

Step 1: Substrate Preparation

Start with a sapphire or silicon wafer. Polish it until the surface is atomically flat—smoother than any mirror you've ever seen. Surface roughness must be below 0.1 nanometers. For context, a human hair is about 100,000 nanometers wide. [3]

Step 2: Metal Deposition

Deposit aluminum or niobium onto the substrate in layers just a few hundred nanometers thick—thinner than a soap bubble. This is done in vacuum chambers using molecular beam epitaxy or sputtering, processes where individual atoms are deposited one layer at a time.[4]

Step 3: Lithography

Pattern the metal using electron-beam lithography—essentially using beams of electrons as a microscopic cookie cutter to carve quantum circuit shapes at nanometer precision. A single mistake, a tiny vibration during exposure, ruins the entire pattern.

Step 4: Etching

Remove unwanted metal through chemical or plasma etching. This is like trimming burnt cake edges, except the "edges" are measured in billionths of a meter and the "trimming" is done with reactive gases.

Step 5: Josephson Junction Creation

This is where the magic happens—and where most fabrication runs fail.

A Josephson junction is a quantum sandwich: two superconducting layers separated by an insulating oxide barrier just 1–2 nanometers thick. That's about 10 atoms.[5]

Creating this barrier requires growing aluminum oxide at exactly the right thickness. Too thick, and electrons can't tunnel through. Too thin, and it's electrically shorted. The tolerance is measured in individual atomic layers.

When it works, electrons tunnel through the barrier—not over it, through it—via quantum tunneling. That tunneling creates the quantum oscillations that form the 0s, 1s, and superpositions that make quantum computing possible.

William Shockley, one of the inventors of the transistor, once said that making reliable semiconductor junctions was "black magic."[6] Josephson junctions are black magic performed at cryogenic temperatures with nanometer precision.

The challenge isn't just creating the junction once in a laboratory. It's creating thousands of identical junctions, consistently, on production timescales. Each Josephson junction must have nearly identical electrical properties for the quantum processor to function correctly. A five percent variation in junction resistance across a chip means five percent variation in qubit frequencies, which means five percent of qubits operating outside their design parameters. This manufacturing challenge, more than any theoretical limitation, is why quantum processor yields remain in the 30–50% for research-grade qubits.[7] And why companies like IBM have built dedicated quantum chip fabrication facilities rather than relying on commercial semiconductor foundries. The good news: yields are improving. For every two chips fabricated, one fails. At $50,000 -100,000 per fabrication run, that failure rate is what keeps quantum hardware teams awake at night.

Quantum Chip Fabrication Process	
Process Step	**What Happens**
Substrate Preparation	Polish wafer to atomic smoothness (<0.1 nm).
Metal Deposition	Deposit aluminum/niobium layers (100–200 nm thick).
Lithography	Pattern circuits using electron beams.
Etching	Remove unwanted metal with reactive gases.
Josephson Junction Creation	Grow 1–2 nm oxide barrier between superconducting layers.
Packaging	Wire and mount chip in dilution refrigerator.

Reading Without Destroying: The Measurement Problem

Remember from Chapter 6: measuring a qubit collapses its quantum state. So how do you read the answer without destroying the calculation?

Indirectly. Very, very carefully.

Each qubit is coupled to a microwave resonator—essentially a tiny electromagnetic cavity that vibrates at a specific frequency. The resonator's frequency shifts slightly depending on whether the qubit is in state $|0\rangle$ or $|1\rangle$.

Send a microwave pulse into the resonator. Measure the reflected signal. The tiny frequency shift—just a few megahertz out of gigahertz—tells you the qubit's state without directly measuring it.[8]

It's like checking if someone's awake by listening for breathing rather than shaking their shoulder. The measurement is gentle enough that it doesn't destroy adjacent qubits still in superposition.

This process, called dispersive readout, works about 97–99% of the time. That 1–3% error rate is actually one of the bigger error sources in current quantum computers.[9]

Control: Conducting the Quantum Orchestra

Qubits don't respond to "run programme.exe." They respond to precisely shaped microwave pulses at gigahertz frequencies—think of them as notes played to choreograph the qubits' quantum state.

Each pulse has three critical parameters:

Frequency: Must match the qubit's resonance, typically 4–8 GHz for superconducting qubits. Off by even a few megahertz, and the pulse does nothing.

Duration: Determines which quantum gate the pulse performs. A $\pi/2$-pulse (half rotation) creates superposition. A π-pulse (full rotation) flips the qubit from $|0\rangle$ to $|1\rangle$.

Phase: The relative timing determines how quantum states interfere with each other.

Getting these parameters right requires calibration. Constant calibration. Daily calibration.

Quantum control is less like programming and more like maintaining an extraordinarily sensitive scientific instrument while simultaneously trying to use it. Qubit frequencies drift as helium levels change in the dilution refrigerator. Temperature fluctuations of mere microKelvin shift resonance frequencies measurably. Stray magnetic fields from nearby electrical equipment, changing throughout the day as usage patterns shift, can detune qubits from their calibrated operating points.

Published calibration protocols from IBM's quantum systems team show that superconducting quantum processors require daily recalibration before use. The calibration sequence characterizes each qubit's current resonance frequency, measures qubit-qubit coupling strengths, and adjusts pulse parameters to compensate for drift since the previous calibration.[2] For a 127-qubit system, this process takes approximately thirty minutes. For larger systems, longer.

This is the operational reality of quantum computing that performance benchmarks don't capture. A quantum processor that achieves 99.5 percent gate fidelity in an ideal calibrated state may perform significantly worse hours later as environmental conditions shift. Maintaining consistent performance requires continuous monitoring, frequent recalibration, and control systems sophisticated enough to track dozens of parameters simultaneously.

Quantum control is less like programming and more like tuning a piano that's constantly going out of tune while you play it.

The Wiring Problem

Here's a problem nobody anticipated when quantum computing was still theoretical: wires.

Each qubit needs roughly three wires. For 100 qubits, that's 300 wires. For 1,000 qubits—the scale needed for truly useful quantum computers—that's 3,000 wires, all threading from room temperature down through the dilution refrigerator to the quantum chip.

Each wire is a thermal path. A highway for heat to leak down to the coldest stage. Current dilution refrigerators can handle maybe 100–200 wires before the cooling power can't keep up.

This isn't a theoretical problem. It's the practical limit preventing quantum computers from scaling beyond a few hundred qubits right now.[11]

The solution? Move control electronics closer to the qubits—inside the refrigerator.

Intel's Horse Ridge II cryogenic control chip, demonstrated in 2022, showed that control electronics could operate at 4 Kelvin while managing multiple qubits simultaneously.[21] Rather than threading hundreds of wires from room temperature to the quantum chip, Horse Ridge allows a small number of high-bandwidth connections to reach a cryogenic controller, which then manages individual qubits locally.

By 2024, multiple companies had released or announced cryogenic control solutions[12]. IBM published results showing integrated cryogenic amplifiers that reduce the signal-to-noise penalty of reading qubit states. Rigetti demonstrated control architectures designed explicitly for modular scaling. The consensus emerging from these efforts is that classical control electronics must move inside the refrigerator to reach qubit counts exceeding a few hundred.[11]

The engineering challenge is significant. Classical electronics generate heat. Heat is lethal to quantum states. Designing electronics that operate at 4 Kelvin while generating minimal heat and managing signals at the precision required for qubit control represents a separate engineering discipline from the qubit physics itself.

It's the quantum equivalent of hiring local managers instead of trying to micromanage from headquarters. Cryogenic control electronics are becoming as important as the qubits themselves.

Five Ways to Build a Qubit

We've focused on superconducting qubits because they're the most mature. But remember from Chapter 7: there are competing approaches, each with different fabrication challenges.

Trapped ion qubits require ultra-high vacuum chambers and laser systems stable to within nanometers. The ions must be trapped using electromagnetic fields with MHz precision while being hit by multiple synchronized laser beams.[13] IonQ's systems use ytterbium-171 ions trapped 50 micrometers apart—about the width of a human hair.

Photonic qubits need single-photon sources that can generate individual photons on demand with >99% purity, then route them through integrated photonic circuits etched into silicon chips.[14]

Xanadu's systems use squeezed light states generated through nonlinear optics in tiny waveguides.

Spin qubits require growing ultra-pure silicon with isotopically enriched silicon-28 (removing Si-29, which has a nuclear spin that causes decoherence), then implanting single phosphorus atoms as qubits and controlling them with microwave pulses and local electric fields.[15]

Each approach has different fabrication challenges. Different failure modes. Different scaling problems.

Nobody knows which will win.

The Scaling Wall

Here's what keeps quantum engineers awake at night: Everyone can build a 2-qubit demonstration. Building 2,000 qubits that actually work together? That's the Mount Everest of quantum engineering.

The challenges compound:

Crosstalk—qubits unintentionally affecting neighbours—is manageable at 100 qubits but becomes a nightmare of unwanted interactions at 1,000[10]. Manufacturing yield drops as chips get larger; achieving the 95% yield needed for reliable systems currently requires multiple fabrication attempts per working chip.[7] And error rates for two-qubit gates—currently around 98.5%—need to reach 99% before error correction becomes practical.[15] Close. Not close enough yet.

Manufacturing yield: Every chip has defects. As chips get larger, yield drops exponentially. Need 95% yield for reliable systems.

Error rates: Single-qubit gates need 99.9% fidelity. Two-qubit gates need 99%. Current state: 99.5% and 98.5%. Close, but not close enough for error correction.[16]

That's why IBM's Heron processor delivers 133 qubits with improved error performance, while their Condor chip reaches 1,121 qubits at the cost of higher error rates.[17] Google's Willow chip targets 105 qubits optimised specifically for error correction. Not because they can't fit more on a chip. Because keeping them all working simultaneously is brutally hard.

The hardware landscape shifted significantly in 2024. IBM released their Heron processor, featuring 133 qubits with a new tunable coupler architecture that reduces unwanted crosstalk between neighbouring qubits. The design addresses one of the primary error sources in superconducting systems: qubits inadvertently affecting their neighbours during operations intended for a single target.

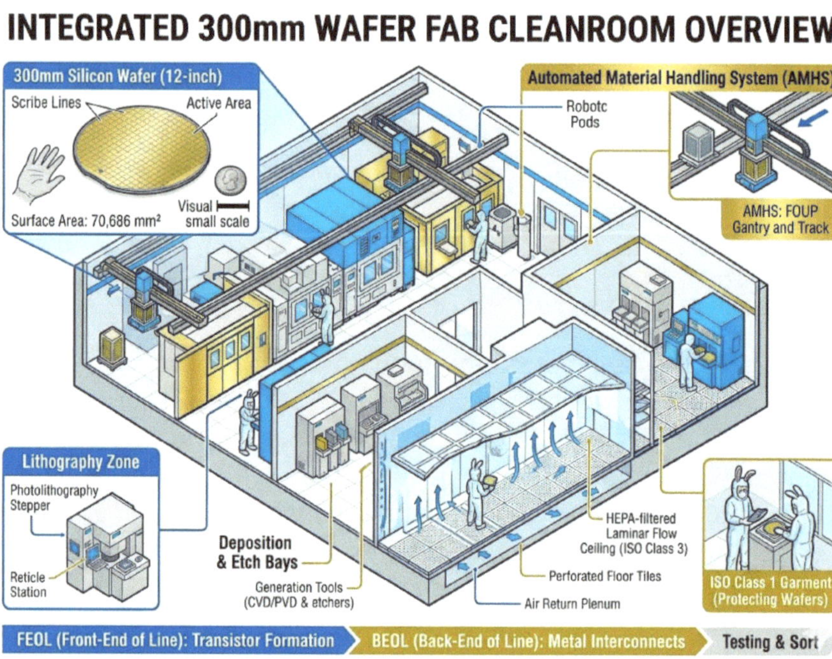

Google's December 2024 Willow chip represented a different advance. Rather than maximizing qubit count, Google optimised for error correction, demonstrating that their surface code implementation could achieve sub-threshold error rates[18], meaning adding more physical qubits actually reduced logical error rates rather than increasing them. They tested this from a 49-qubit to a 105-qubit configuration, observing tenfold error reduction with each doubling.

Perhaps most surprisingly, Atom Computing announced a 1,180-qubit neutral atom system using a completely different fabrication approach.[20] Instead of superconducting circuits manufactured in cleanrooms, they use individual ytterbium atoms held in place by focused laser beams. The qubits are the atoms themselves, not engineered circuits. Reconfiguring the system means rearranging the laser pattern rather than building a new chip.

These three announcements in a single year illustrate both the pace of progress and the diversity of approaches still competing for dominance. The fabrication challenges are different for each technology, the scaling strategies are different, and the performance tradeoffs are different. We are still early enough in quantum hardware development that the winning approach or approaches remain genuinely uncertain.[17]

The Modular Future

The new strategy: stop trying to build one giant chip. Build modules.

Instead of 1,000 qubits on one chip, build ten modules of 100 qubits each. Connect them via photonic links—essentially using light to move quantum information between modules.[19]

If it works, you could build cloud-based quantum data centres. Each refrigerator hosts a quantum module. String them together through fiber optic connections. Scale horizontally instead of vertically.

Key Components of a Quantum Computer

Component	Function
Qubit Chip	Contains superconducting circuits that encode and process quantum information.
Cryostat	Maintains temperature at ~15 mK through dilution refrigeration.
Control Lines	Carry microwave pulses from room-temperature electronics to manipulate qubit states.
Readout Resonators	Measure qubit states without collapsing neighbouring qubits.
Classical Backend	Generates control signals, processes measurement results, and manages calibration.

Harvard, MIT, and several companies are pursuing this. It might be how we actually get to millions of qubits: not one massive chip, but thousands of small, reliable modules working together.

Or it might not work at all. That's quantum hardware.

Quantum hardware is the ultimate blend of precision engineering and controlled chaos. It's like building a spaceship that only flies at temperatures colder than deep space, controlled by wires thinner than hair, performing calculations that collapse if you breathe too hard nearby.

And yet, every year, engineers squeeze more qubits, longer coherence times, and higher fidelity from materials that shouldn't work at all.

One day—maybe in a decade, maybe two—you might have quantum co-processors in data centres solving optimization problems that classical computers can't touch. You won't see the dilution refrigerator. You won't hear the microwave pulses. You'll just submit a problem and get an answer that was computationally impossible yesterday.

But right now? Right now it's golden chandeliers, careful wire routing, daily calibration, and engineers celebrating when they get 100 microseconds of coherence time instead of 50.

That's where we are. Quantum hardware is still hard. Magnificently, frustratingly, expensively hard.

But possible. Barely. And getting better.

Turn the page.

References

[1] Krinner, S., et al. (2019). Engineering cryogenic setups for 100-qubit scale superconducting circuit systems. EPJ Quantum Technology, 6(2).

[2] Krantz, P., et al. (2019). A quantum engineer›s guide to superconducting qubits. Applied Physics Reviews, 6(2), 021318.

[3] Oliver, W. D., & Welander, P. B. (2013). Materials in superconducting quantum bits. MRS Bulletin, 38(10), 816–825.

[4] Devoret, M. H., & Schoelkopf, R. J. (2013). Superconducting circuits for quantum information: An outlook. Science, 339(6124), 1169–1174.

[5] Clarke, J., & Wilhelm, F. K. (2008). Superconducting quantum bits. Nature, 453(7198), 1031–1042.

[6] Riordan, M., & Hoddeson, L. (1997). Crystal fire: The invention of the transistor. W. W. Norton & Company.

[7] Gambetta, J. M., Chow, J. M., & Steffen, M. (2017). Building logical qubits in a superconducting quantum computing system. npj Quantum Information, 3(2).

[8] Blais, A., et al. (2021). Circuit quantum electrodynamics. Reviews of Modern Physics, 93(2), 025005.

[9] Walter, T., et al. (2017). Rapid high-fidelity single-shot dispersive readout of superconducting qubits. Physical Review Applied, 7(5), 054020.

[10] Sheldon, S., et al. (2016). Procedure for systematically tuning up cross-talk in the cross-resonance gate. Physical Review A, 93(6), 060302.

[11] Van Dijk, J. P. G., et al. (2019). The impact of classical control electronics on qubit fidelity. Physical Review Applied, 12(4), 044054.

[12] Patra, B., et al. (2018). Cryo-CMOS circuits and systems for quantum computing applications. IEEE Journal of Solid-State Circuits, 53(1), 309–321.

[13] Bruzewicz, C. D., et al. (2019). Trapped-ion quantum computing: Progress and challenges. Applied Physics Reviews, 6(2), 021314.

[14] Slussarenko, S., & Pryde, G. J. (2019). Photonic quantum information processing: A concise review. Applied Physics Reviews, 6(4), 041303.

[15] Veldhorst, M., et al. (2015). A two-qubit logic gate in silicon. Nature, 526(7573), 410–414.

[16] Acharya, R., et al. (2023). Suppressing quantum errors by scaling a surface code logical qubit. Nature, 614(7949), 676–681.

[17] IBM Quantum. (2024). IBM Quantum Heron processor. IBM Research Blog. https://research.ibm.com/blog/ibm-quantum-heron

[18] Google Quantum AI & Collaborators. (2024). Quantum error correction below the surface code threshold. Nature, 614(7949), 676–681.

[19] Monroe, C., et al. (2014). Large-scale modular quantum-computer architecture with atomic memory and photonic interconnects. Physical Review A, 89(2), 022317.

[20] Bluvstein, D., et al. (2024). Logical quantum processor based on reconfigurable atom arrays. Nature, 626(7997), 58–65.

[21] Intel Corporation. (2022, February). Intel introduces Horse Ridge II to enable control of multiple qubits [Press release]. Intel Newsroom.

CHAPTER

9

How to Talk to a Qubit

The software stack that connects to quantum mechanics

Wait—quantum computers have software?

Yes. Shocking, I know.

Contrary to what science fiction suggests, quantum computers aren't mysterious black boxes humming in alien code. They have software stacks—layered systems that translate human-friendly Python into quantum gates, then into microwave pulses, then into actual physical qubit rotations.

> *Think of it as the world's most demanding translation job. You write something recognizable—Python code on a laptop—and by the time your instruction reaches its destination, it has been translated through four completely different languages.*

Each more alien than the last, until finally a precisely shaped microwave pulse lasting thirty nanoseconds nudges a superconducting circuit colder than outer space into doing what you originally asked.

Most of the time it works. Sort of. Approximately.

The quantum software stack has four layers, each translating your intention into something the next layer understands.

At the top, you write Python code. The Compiler Layer below translates it into quantum gates. The Pulse Layer converts gates into precisely timed microwave signals. At the bottom, the Hardware

Layer—the qubits—receive those signals and change their quantum states.

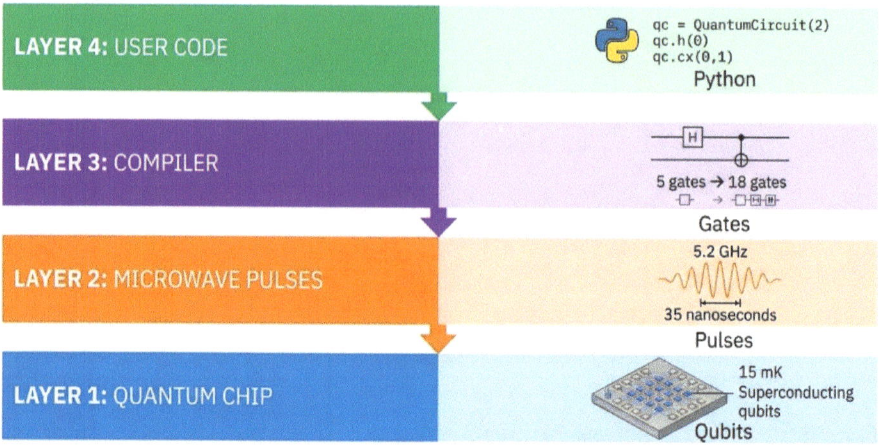

Four layers translate Python code into quantum operations: User writes code (green) → Compiler converts to gates (purple) → Pulses control qubits (orange) → Hardware executes (blue).

Layer 1: Where Humans Live

At the top of the stack, you write code in something that looks almost normal. Almost.

IBM's Qiskit framework lets you write quantum circuits in Python.[1,5] Google's Cirq does the same.[2] Microsoft has Q#, a language that looks like C# but thinks in quantum states—precise, strongly typed, and genuinely useful once you accept that programming here means manipulating probability amplitudes rather than variables.[3]

Here's a complete quantum programme:

```
from qiskit import QuantumCircuit
qc = QuantumCircuit(1, 1)
qc.h(0)            # Hadamard gate - creates superposition
qc.measure(0, 0) # Measure the qubit
```

This isn't pseudocode. This runs. Right now, on IBM's quantum computers, accessible free through their cloud platform.

> *The qubit that executes this programme is a superconducting circuit in Yorktown Heights, New York, cooled to fifteen millikelvin, maintained by engineers who spend their days fighting decoherence. And you just told it what to do from your laptop.*

That's genuinely remarkable, even if the result is just a coin flip.

Run this 1,000 times, and you'll get roughly 500 zeros and 500 ones—a quantum coin flip.

Except unlike a classical coin flip, where the coin is always heads or tails (you just don't know which), the qubit is genuinely both until you measure it. The superposition is real. The measurement forces it to choose.

That philosophical difference took physicists thirty years of arguments to accept. Your Python code just does it in one line.

Layer 2: The Compiler (Quantum GPS)

Your four-line programme looks simple. But here's what happens behind the scenes.

IBM calls their quantum compiler the Transpiler. It takes your abstract circuit and transforms it into something the actual hardware can execute.

Problem: You wrote a CNOT gate between qubit 0 and qubit 5. But on IBM's quantum computer, qubit 0 isn't physically connected to qubit 5. It's only connected to qubits 1 and 14.

The compiler needs to route your operation through intermediate qubits using SWAP gates. Your elegant two-qubit gate becomes a sequence of seven gates using qubits 0, 1, 14, and 5.

Every extra gate adds error. Your code said "do this operation with 99% fidelity." The compiler's version might have 93% fidelity because of all the routing overhead introduced by limited qubit connectivity.

This is like asking your GPS for directions and having it say, "Sure, but construction closed your direct route, so I'll send you through three detours, adding 40 minutes and a toll booth."

Welcome to quantum compilation.

IBM's compiler team—which they call the Transpiler team—has published extensively on this problem. Their research shows that a quantum circuit requiring 100 logical gates can easily expand to 400 physical gates after routing through hardware connectivity constraints.[4] Each additional gate adds noise. Each routing detour degrades fidelity.

The mathematics is unforgiving. If each gate has a 0.5% error rate, then 100 gates give you roughly a 60% chance of a clean result. Four hundred gates after compilation? Your probability drops below 15%.

This is why compiler optimization in quantum computing isn't a convenience feature. It's the difference between getting an answer and getting noise.

That's quantum software development: managing disappointment at scale.

Layer 3: Pulse Control (The Concert)

After compilation, your quantum gates need to become physical operations on actual qubits.

For superconducting qubits, this means microwave pulses at 4–8 GHz, carefully shaped with Gaussian envelopes, precise durations of 20–40 nanoseconds, and phases controlled to within fractions of a degree.[6]

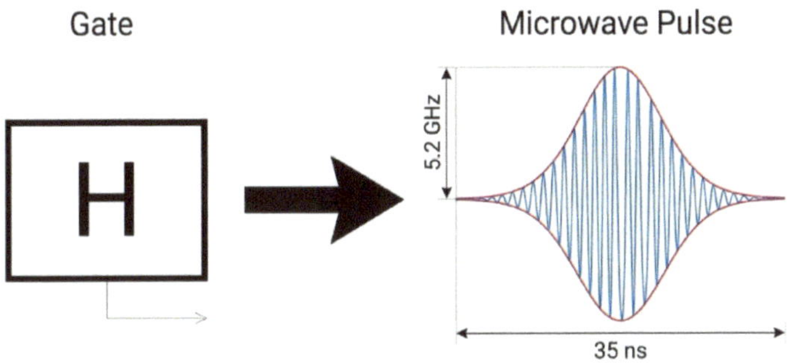

A quantum gate becomes a precisely shaped microwave pulse:
Hadamard gate (H) → 35-nanosecond Gaussian pulse at 5.2 GHz.

IBM's OpenPulse lets advanced users write these pulses directly.[7] You specify the waveform, frequency, amplitude, and duration.

It's like editing a film where every cut must land within 2 nanoseconds—miss the timing and the entire scene is noise, not signal.

Most people don't use OpenPulse. It's too low-level, too easy to break things. But for researchers squeezing every last bit of performance from quantum hardware, it's essential.

The pulse layer is where quantum software becomes quantum jazz—improvisation constrained by physics.

Let's Actually Write Some Quantum Code

Here's a classical algorithm searching for a specific number:

```
# Classical search (check each item)
numbers = [0, 1, 2, 3, 4, 5, 6, 7]
target = 5
for i, num in enumerate(numbers):
    if num == target:
        print(f"Found it at position {i}!")
```

This checks eight numbers one at a time. Eight operations.

Now here's Grover's quantum search algorithm doing the same thing:

```
from qiskit import QuantumCircuit
qc = QuantumCircuit(3, 3)

# Put all qubits in superposition (all 8 numbers simultaneously)
qc.h([0, 1, 2])

# Grover operator (marks the target, amplifies its probability)
qc.cz(0, 1)  # Oracle marks |5⟩ = |101⟩
qc.cz(1, 2)

# Diffusion operator (amplifies marked state)
qc.h([0, 1, 2])
qc.x([0, 1, 2])
qc.h(2)
qc.ccx(0, 1, 2)
qc.h(2)
qc.x([0, 1, 2])
qc.h([0, 1, 2])

qc.measure([0, 1, 2], [0, 1, 2])
```

This checks all eight numbers simultaneously through quantum superposition, then amplifies the probability of measuring the correct answer.[8]

Run it, and you'll measure |101⟩ (which is 5 in binary) with high probability.

Classical: Eight sequential checks.

Quantum: A single pass using superposition and interference to make the correct answer the most probable result.

The difference looks modest for eight items. For a database with a billion entries, Grover's algorithm needs roughly 31,623 operations instead of a billion.[9] For a trillion entries, about a million operations instead of a trillion. The speedup isn't linear—it's quadratic, scaling with the square root of the problem size.

That's not enough to break encryption on its own. But for search problems—finding a needle in an enormous haystack—it's the difference between feasible and impossible.

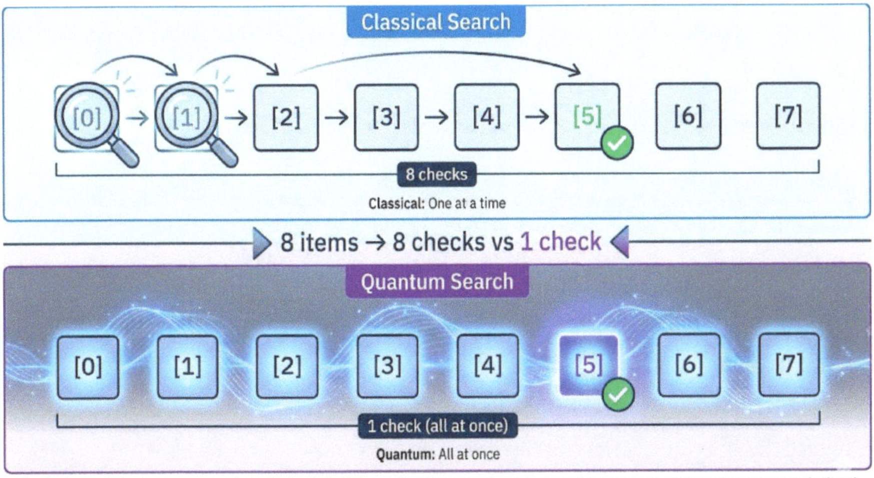

Classical search checks items sequentially (8 checks for 8 items). Quantum search uses superposition to check all items simultaneously (1 check).

The Quantum Languages

There are now half a dozen serious quantum programming frameworks, each with different philosophies:

Qiskit (IBM): Python-based, huge community, runs on real IBM quantum hardware. If you're learning quantum programming, start here.[10]

Cirq (Google): Also Python, more flexible for research, integrates with TensorFlow for quantum machine learning.[11]

Q# (Microsoft): Looks like C#, strongly typed, designed for eventual fault-tolerant quantum computers that don't exist yet.[12]

PennyLane (Xanadu): Python framework specifically for quantum machine learning, works with any quantum hardware backend.[13]

Braket SDK (Amazon): AWS-integrated, lets you run quantum code on IonQ, Rigetti, or D-Wave hardware through Amazon's cloud.[14]

It's like choosing between spoken languages. They all let you communicate with qubits. Some are more verbose (Q#). Some are more flexible (Cirq). Some have bigger communities (Qiskit).

The good news: they're all free. The bad news: learning one doesn't mean you know the others. Quantum languages haven't standardized yet.

Each quantum framework comes with a *personality.*

Framework	Best For	Primary Platform
Qiskit (IBM)	General purpose, education, research	IBM Quantum
Cirq (Google)	Research, custom circuit design	Google quantum hardware
Q# (Microsoft)	Enterprise, fault-tolerant design	Azure Quantum
PennyLane (Xanadu)	Quantum ML, hybrid ML-quantum models	Hardware-agnostic
Braket SDK (Amazon)	Cloud access to multiple quantum devices	IonQ, Rigetti, D-Wave

We're in the 1960s of quantum computing, when every computer manufacturer had their own assembly language. Eventually, standards will emerge. Until then, it's fragmented.

The Quantum Hello World

Every programming language has a "Hello, World!" programme. In quantum computing, it's the quantum coin flip:

```
from qiskit import QuantumCircuit, Aer, execute

# Create circuit with 1 qubit and 1 classical bit
qc = QuantumCircuit(1, 1)

# Apply Hadamard gate (creates superposition)
qc.h(0)

# Measure the qubit
qc.measure(0, 0)

# Run on simulator
backend = Aer.get_backend('qasm_simulator')
result = execute(qc, backend, shots=1000).result()
print(result.get_counts())
```

Output: `{'0': 498, '1': 502}`

Congratulations. You've created a quantum superposition and measured it 1,000 times.

Roughly half came up 0, half came up 1. This is the simplest possible quantum programme. One qubit. One gate. One measurement. And yet it demonstrates superposition, measurement collapse, and probabilistic results—the three fundamental quantum quirks of quantum mechanics.

Frameworks like Pennylane and Qiskit Runtime are built for this partnership—blending quantum acceleration into everyday Python workflows.

Running on Real Quantum Hardware

Here's how you actually run a quantum programme on IBM's hardware:

Step 1: Create a free IBM Quantum account.[15] They give you access to real quantum computers—not simulators, actual superconducting qubits in dilution refrigerators in Yorktown Heights, New York.

Step 2: Write your code in Qiskit.

Step 3: Simulate it locally first. The simulator shows you what would happen in a perfect, noise-free world. If your code doesn't work in the simulator, it definitely won't work on real hardware.

Step 4: Choose a backend. IBM's available systems change frequently—check quantum.ibm.com/services/resources for the current list of machines and qubit counts.

Step 5: Submit your job to the queue.

Step 6: Wait. Sometimes minutes. Sometimes hours. Depends on how many other people are using that quantum computer.

Step 7: Get your results. They'll be noisy, probabilistic, and possibly wrong. That's normal.

Step 8: Run it 1,000 more times and average the results. Quantum computing is statistical.

The first time you see your code actually execute on a real quantum computer—watching your qubits rotate on an actual superconducting chip colder than outer space—is genuinely magical.

Then you realize it gave you the wrong answer because of decoherence, and the magic fades slightly.

But still. You just programmed quantum mechanics. That's worth something.

Why Your Quantum Programme is Probably Wrong

Here's what nobody tells beginners: quantum programmes fail constantly.

Your five-line quantum circuit becomes 50 physical gates after compilation. Each gate has a 1% error rate. Errors compound. By the end, you're getting noise, not signal.

Published documentation from IBM's quantum team describes this adjustment consistently. Quantum computing is inherently probabilistic. A circuit that should produce a specific answer produces that answer with some probability, along with incorrect answers with lower probabilities. The noise floor—the minimum error rate from decoherence, gate infidelity, and measurement errors—means no quantum calculation is ever perfectly reliable.[16]

The practical response is statistical. Run the same circuit thousands of times. Plot the distribution of outcomes. The correct answer typically appears as a clear peak above the noise, identifiable not because it's always right but because it's right more often than anything else.

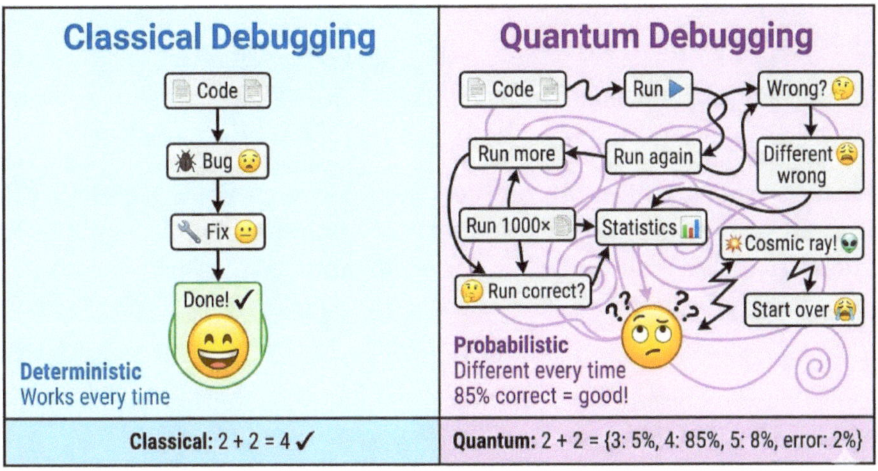

Classical debugging: Find bug, fix it, done. **Quantum debugging:** Run 1,000 times, analyze probability distribution, accept 85% accuracy as success.

This is fundamentally different from debugging classical programmes. You're not finding the line where the logic breaks. In classical code, 2 + 2 = 4 every time. In quantum code, 2 + 2 might give you 4 with 85% probability, 3 with 10% probability, and complete garbage 5% of the time. You're characterizing a probability distribution and deciding whether the signal is strong enough to trust.

It's less like engineering and more like experimental physics.

The Hybrid Future

Pure quantum computing is years away. What works today is hybrid quantum-classical computing.[17]

The workflow:

1. Classical computer preprocesses the problem, simplifies it, identifies the quantum-suitable parts.
2. Quantum computer runs a short circuit (before decoherence destroys everything).
3. Classical computer analyzes results, potentially feeds them back into another quantum iteration.

This is how quantum machine learning works. Classical neural networks do most of the work. Quantum circuits handle specific optimization steps where quantum parallelism helps.[18]

IBM's Qiskit Runtime, Google's Cirq with TensorFlow Quantum, Amazon's Braket—they're all built around this hybrid model.[19]

Because here's reality: for the next decade, quantum computers will be coprocessors, not replacements. They'll handle small, specific tasks where quantum mechanics provides advantage. Classical computers will do everything else.

It's like GPUs for graphics rendering. CPUs handle the game logic. GPUs handle the intensive parallel calculations. Neither replaces the other.

Quantum processors will eventually be the same—specialized accelerators for specific computational tasks.

The Golden Rule of Quantum Coding

> *"If it runs perfectly the first time, it's probably classical."*

Quantum programming is about experimentation, observation, and learning through probabilistic outcomes.

The Quantum Developer's Toolset

Category	Tool	Purpose
IDE	Qiskit Notebook/Jupyter	Visual, step-by-step development
Simulator	Aer (Qiskit), Cirq Simulator	Noise-free and noisy circuit testing
Cloud Access	IBM Quantum, AWS Braket, Azure Quantum	Access to real quantum hardware
Visualization	Bloch Sphere, plot_histogram	Visualize qubit states and measurement results
Debugging	Qiskit Pulse, Transpiler logs	Low-level pulse control and optimization tracking

What This All Means

Fifteen years ago, quantum computing was pure theory discussed in academic journals. Ten years ago, you needed a physics PhD and institutional access to touch quantum hardware. Five years ago, you needed a research budget. Today, you need a laptop, an internet connection, and twenty minutes to create an IBM Quantum account.

That trajectory is remarkable. The software infrastructure is real, functional, and free to access. The hardware is real, operational, and available in the cloud. The algorithms are real, documented, and teachable.

What's missing is scale. Current quantum computers are too small, too noisy, and too limited for the applications that would justify their existence in every data centre. The software is ready. The physics is understood. The engineering needs another decade.

But you can start learning today. Write a quantum circuit. Submit it to a real quantum computer. Watch probability distributions emerge from quantum mechanics. The tools exist. The access exists.

The hardware will catch up.

But while it does—while engineers fight decoherence one microsecond at a time—the real race is happening somewhere else entirely.

Not in the lab. In the boardrooms, defence ministries, and intelligence agencies of the world's superpowers.

Chapter 10.

References

[1] IBM Quantum. (2024). Qiskit documentation. https://qiskit.org/documentation/

[2] Google Quantum AI. (2024). Cirq: A Python framework for quantum circuits. https://quantumai.google/cirq

[3] Microsoft Azure Quantum. (2024). Q# language documentation. https://learn.microsoft.com/azure/quantum/user-guide/

[4] McKay, D. C., Wood, C. J., Sheldon, S., Chow, J. M., & Gambetta, J. M. (2017). Efficient Z-gates for quantum computing. *Physical Review A*, *96*(2), 022330.

[5] Javadi-Abhari, A., Treinish, M., Krsulich, K., Wood, C. J., Lishman, J., Gacon, J., Martiel, S., Nation, P. D., Bishop, L. S., Cross, A. W., Johnson, B. R., & Gambetta, J. M. (2024). Quantum computing with Qiskit. arXiv:2405.08810.

[6] Krantz, P., et al. (2019). A quantum engineer's guide to superconducting qubits. *Applied Physics Reviews*, *6*(2), 021318.

[7] Alexander, T., Kanazawa, N., Egger, D. J., Capelluto, L., Wood, C. J., Javadi-Abhari, A., & McKay, D. C. (2020). Qiskit Pulse: Programming quantum computers through the cloud with pulses. *Quantum Science and Technology*, *5*(4), 044006.

[8] Grover, L. K. (1996). A fast quantum mechanical algorithm for database search. *Proceedings of the 28th Annual ACM Symposium on Theory of Computing*, 212–219.

[9] Boyer, M., Brassard, G., Høyer, P., & Tapp, A. (1998). Tight bounds on quantum searching. *Fortschritte der Physik*, *46*(4–5), 493–505.

[10] Qiskit Community. (2024). Learn quantum computation using Qiskit. https://qiskit.org/learn/

[11] Cirq Developers. (2024). Cirq (Version 1.3.0). https://quantumai.google/cirq

[12] Microsoft Quantum Team. (2024). Introduction to Q# and the Quantum Development Kit. https://learn.microsoft.com/azure/quantum/overview-what-is-qsharp-and-qdk

[13] Bergholm, V., Izaac, J., Schuld, M., Gogolin, C., Ahmed, S., Ajith, V., Alam, M. S., Alonso-Linaje, G., AkashNarayanan, B., Asadi, A., Arrazola, J. M., Azadian, U., Banning, S., Blank, C., Bromley, T. R., Cordier, B. A., Ceroni, J., Delgado, A., Di Matteo, O.,... Killoran, N. (2022). PennyLane: Automatic differentiation of hybrid quantum-classical computations. arXiv:1811.04968.

[14] Amazon Web Services. (2024). Amazon Braket developer guide. https://docs.aws.amazon.com/braket/

[15] IBM Quantum. (2024). IBM Quantum Experience. https://quantum.ibm.com/

[16] Preskill, J. (2018). Quantum computing in the NISQ era and beyond. *Quantum, 2*, 79.

[17] Benedetti, M., Lloyd, E., Sack, S., & Fiorentini, M. (2019). Parameterized quantum circuits as machine learning models. *Quantum Science and Technology*, *4*(4), 043001.

[18] Biamonte, J., Wittek, P., Pancotti, N., Rebentrost, P., Wiebe, N., & Lloyd, S. (2017). Quantum machine learning. *Nature, 549*(7671), 195–202.

[19] IBM Quantum. (2024). Qiskit Runtime documentation. https://qiskit.org/ecosystem/ibm-runtime/

CHAPTER

10

The Two Paths to Quantum Computing

One calculates. The other relaxes.

Imagine two people trying to reach the summit of Mount Everest.

The first hires Sherpa guides, studies weather windows, trains for a year, and picks the technical route up the Khumbu Icefall. Every step is deliberate. Every decision calculated. The summit is possible—but only if everything goes right.

The second looks at the mountain for a long moment, then says: "I'm going to let gravity decide. I'll find the path of least resistance and follow it down until I reach the lowest point."

Different goals. Different methods. Same mountain.

That's gate-model quantum computing versus quantum annealing. Both use the same quantum mechanics—superposition, tunneling, entanglement. Both are trying to solve problems that defeat classical computers. But they think about problems in fundamentally different ways, and they're suited to completely different tasks.

One calculates. The other relaxes.

The Gate Model: Playing Quantum Chess

> *Gate-model quantum computers are strategic thinkers. They manipulate qubits step by step using quantum gates—mathematical operations that rotate quantum states in precise ways.*

String enough gates together, and you've got a quantum circuit that solves a problem.

Each gate is a move. The circuit is the strategy. The measurement is checkmate.

> *Running Shor's algorithm to factor a 2,048-bit encryption key—the kind protecting your bank account—requires roughly 10 million quantum gates arranged in a precise sequence.*[1]

Each gate must execute with better than 99.9% accuracy. The entire calculation must finish before decoherence collapses the quantum state back to classical noise. Miss either requirement, and the answer is garbage.

That's why IBM and Google have spent billions building superconducting qubits that can execute gates in 20–40 nanoseconds with high precision.[2] Every nanosecond matters. Every gate must be perfect.

The gate model is universal quantum computation—given enough qubits and low enough error rates, it can run any quantum algorithm ever devised. Shor's algorithm. Grover's search. Quantum simulation. Machine learning. Chemistry. Cryptography. Everything.[3]

But that universality comes at a cost: extreme sensitivity to noise, massive calibration requirements, and the need for error correction that multiplies qubit counts by factors of 1,000.

Quantum Annealing: The Zen Garden Approach

Now imagine instead of playing chess, you're trying to find the lowest point in a mountain range—the deepest valley where water would naturally collect.

A quantum annealer starts with all possibilities simultaneously (mountains, hills, valleys, plains) and slowly, gradually lets the system relax until the solution naturally settles into the lowest-energy configuration.

That lowest-energy state is your answer.

While the gate model calculates with logic, the annealer lets physics do the math.

Here's how it works: Every optimization problem can be represented as an energy landscape. Good solutions are low-energy valleys. Bad solutions are high-energy peaks. The optimal solution is the global minimum—the deepest valley.

A quantum annealer encodes your problem into this landscape using qubits and couplings between them. Then it slowly—adiabatically is the physics word for it—evolves the system from a simple starting state toward the complex problem state. Think of slowly turning up the difficulty while the system continuously adjusts to find its lowest-energy configuration.[4]

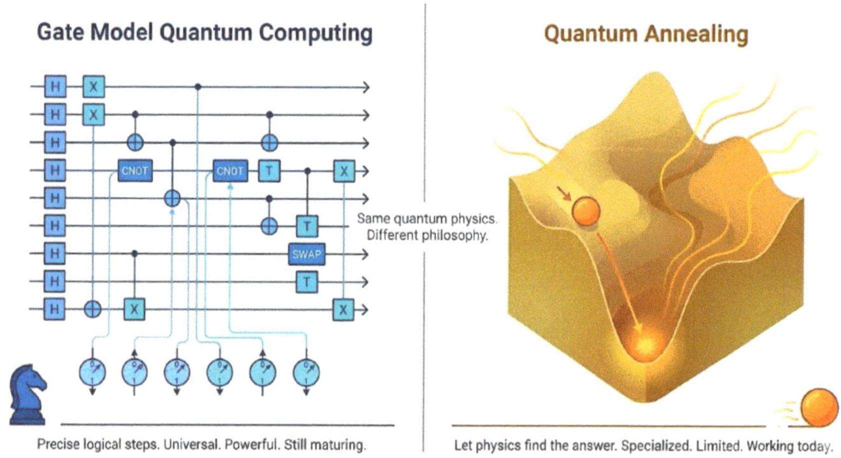

If you do it slowly enough, quantum mechanics guarantees the system stays in its lowest energy state throughout. Start with the ground state of a simple problem, evolve to a hard problem, and you end up with the ground state of the hard problem—which is your answer.

The Secret Weapon: Quantum Tunneling

Here's where annealers have an advantage over classical optimization.

Imagine a marble trapped in a shallow valley next to a much deeper one. To get to the better solution, it needs to roll up and over the hill separating them.

Classical physics says: "Sorry, you need enough thermal energy to climb that hill. If you don't have it, you're stuck."

Classical optimization algorithms behave like a marble rolling across hilly terrain. They find a low point and stop, assuming it's the lowest point available. But if there's a deeper valley just beyond a ridge, the marble is stuck. Getting there requires climbing up first—and the algorithm has no incentive to go uphill before going down.

Quantum tunneling changes this. Instead of climbing over the ridge to reach the deeper valley, the quantum annealer's qubits can pass through the barrier entirely.[5] Not around it. Through it. The quantum state penetrates energy barriers that classical physics treats as impassable walls.

This is why quantum annealing is particularly interesting for optimization problems with complex energy landscapes—problems where classical algorithms consistently get trapped in good-but-not-great solutions.

Does It Actually Work?

That's the billion-dollar question. Literally. D-Wave has raised over $200 million betting quantum annealing provides genuine speedup for optimization problems.[6]

The evidence is... complicated.

In 2014, researchers at Google, NASA, and USC tested D-Wave's system against a specific classical algorithm called simulated annealing. On certain carefully chosen problems, quantum annealing was up to 100 million times faster.[7]

The catch—and there's always a catch in quantum computing—was that simulated annealing wasn't the best classical approach for those problems. When researchers used better classical algorithms, the quantum advantage shrank or disappeared entirely.

Then other researchers showed you could beat D-Wave with better classical algorithms. Then D-Wave released newer hardware that beat those algorithms. Then researchers improved classical methods again.[8]

It's an ongoing arms race. What's clear is this: quantum annealing works. Whether it provides practical quantum advantage for real-world problems depends heavily on the problem, the implementation, and what classical algorithm you're comparing against.

But here's what matters: companies are using it. Not for marketing. For actual work.

Volkswagen used D-Wave's annealer to optimise traffic flow in Beijing—routing taxis to minimize congestion.[9] Denso, a major auto parts supplier, used it for factory optimization.[10] Research teams including those at Los Alamos National Lab have used D-Wave systems for quantum-assisted machine learning.[11]

These aren't research papers. They're production deployments where the quantum annealer provided value worth its multimillion-dollar price tag.

The Traveling Salesman Walks into a Quantum Computer

Here's the classic optimization problem: A delivery driver must visit 20 cities and return home. What's the shortest possible route?

For 20 cities, there are $20! = 2.4 \times 10^{18}$ possible routes—that's roughly 2.4 quintillion.[12] Even checking one billion routes per second, a classical computer would take 77 years to try them all. Your delivery driver would retire several times before getting an optimal route.

A quantum annealer doesn't try them all. It encodes the problem as an energy landscape where each route is a different energy state. The shortest route is the lowest-energy state. Then it lets physics find that state through adiabatic evolution and quantum tunneling.

Will it find the absolute optimal route? Maybe. Maybe not. But it will find a very good route very quickly—often good enough for practical logistics where "near-optimal in seconds" beats "optimal in decades."

Gate-model quantum computers could theoretically solve this faster using Grover's algorithm, which provides quadratic speedup.[13] But that requires thousands of logical qubits with error correction—technology that won't exist for another decade.

Annealers work today. Imperfectly. Noisily. But they work.

The Numbers: How They Actually Compare

Let's be quantitative about this:

GATE MODEL QUANTUM SYSTEMS (EARLY 2026)		
System	**Qubits**	**Key Feature**
IBM Heron	133	Tunable-coupler architecture
IBM Condor	1,121	High qubit count
Google Willow	105	Error-correction optimised
Atom Computing	1,180	Neutral-atom based architecture

Gate fidelity: 99.5% single-qubit, 99% two-qubit. Coherence: 100–150 μs. Cost: $10–50 million per system.

QUANTUM ANNEALING SYSTEMS (EARLY 2026)		
System	**Qubits**	**Notes**
D-Wave Advantage	5,000+	Specialized for optimization problems

Annealing time: microseconds to milliseconds. Cost: $10–15 million per system or cloud access.

> *The qubit count comparison is misleading. A D-Wave qubit isn't equivalent to an IBM qubit. It's like comparing a calculator key to a computer CPU transistor—both process information, but one is far more versatile.*

But for the specific problems annealers are designed for, that 5,000-qubit advantage matters.

Hybrid Approaches: Best of Both Worlds

D-Wave's modern systems aren't pure quantum annealers anymore. They're hybrid systems combining classical preprocessing with quantum annealing.[16]

Here's the workflow:

Step 1: Classical computer analyzes the problem, identifies structure, reduces complexity where possible.

Step 2: Problem gets encoded into the quantum annealer's qubit connectivity graph.

Step 3: Quantum annealing runs, producing multiple samples (because quantum results are probabilistic).

Step 4: Classical computer analyzes results, potentially feeds them back to refine the quantum annealing run.

It's like using Google Maps to navigate 90% of your route, then quantum optimization to find shortcuts in real-time.

Gate-model systems are moving this direction too. IBM's Qiskit Runtime does classical-quantum iteration.[18] Google's Quantum AI runs hybrid algorithms.

This hybrid approach has become the dominant paradigm for near-term quantum computing. Neither purely classical nor purely quantum, hybrid algorithms extract whatever advantage current quantum hardware can provide while relying on classical systems to handle the rest.

The terminology the industry now uses is telling: "quantum-centric supercomputing." Rather than replacing classical computers, quantum processors augment them—handling the specific calculations where quantum provides advantage while classical systems manage everything else. IBM's 2024 roadmap explicitly frames their quantum systems as components in hybrid classical-quantum workflows rather than standalone replacements.[17]

This matters for setting realistic expectations. The era of quantum computers sitting in isolation and solving problems independently is probably decades away. The near-term future is quantum processors as specialized accelerators—fast at certain tasks, useless for others, always working alongside classical systems.

The future isn't pure quantum or pure classical. It's intelligent orchestration between them.

When to Use Which

So when should you actually use each approach?

If your problem involves optimization—routing trucks, scheduling staff, allocating financial portfolios, designing supply chains—and you need workable answers today rather than theoretically perfect answers in a decade, quantum annealing is worth investigating. D-Wave's systems are commercial products with documented deployments. They're expensive, specialized, and not right for most problems. But for that specific category of combinatorial optimization, they represent the most mature quantum computing technology currently available.

If your problem involves simulating quantum systems— designing new drug molecules, modeling chemical reactions, understanding materials at the quantum level—gate-model systems are what you want, though you'll probably be waiting years for hardware capable enough to outperform classical simulation. IBM's cloud access lets researchers experiment now with the understanding that truly useful quantum chemistry simulation requires hardware improvements still in progress.

If you're a cryptographer concerned about quantum computers eventually breaking RSA encryption, you should already be migrating to post-quantum cryptographic standards regardless of which quantum architecture wins. NIST published those standards in 2024.[21] The timeline for cryptographically relevant quantum computers is uncertain, but the eventual threat is not.

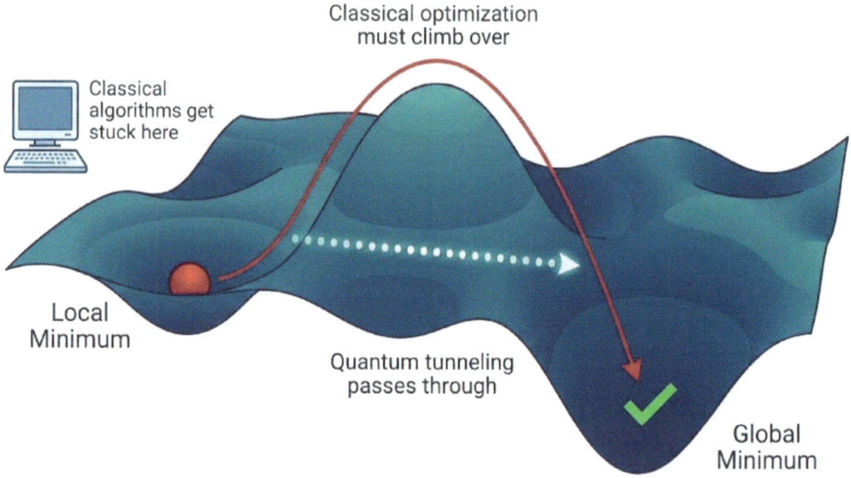

And if someone tells you quantum computing will transform your business within the next two years? Ask them which specific algorithm, which specific problem, and which specific hardware platform. The answer will tell you whether they understand the field or whether they're selling you something.

The QAOA Middle Ground

There is a third approach worth noting: the Quantum Approximate Optimization Algorithm (QAOA), which runs on gate-model hardware but targets the same combinatorial problems as annealers.[19] Early commercial pilots—Volkswagen for traffic optimisation, Airbus for aircraft loading—show promise, though consistent advantage over classical methods remains problem-dependent.

Think of QAOA as the quantum computing equivalent of a hybrid car. Not fully quantum, not fully classical, but potentially the best practical solution while we wait for fault-tolerant quantum systems. Early results are promising—QAOA has demonstrated advantages over classical heuristics on specific graph problems—but the hardware requirements to achieve meaningful advantage are still beyond what current systems reliably deliver.[19]

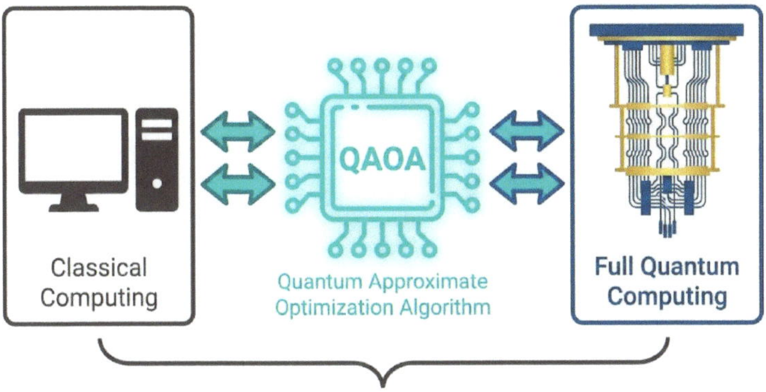

The honest assessment from researchers: QAOA is more interesting than quantum annealing from a theoretical perspective because it runs on universal gate-model hardware and could in principle scale to harder problems. But practical demonstrations of genuine quantum advantage using QAOA remain elusive. It's a promising direction that requires better hardware before its potential can be properly evaluated.

Real Companies, Real Money, Real Results

Gate Model:

- IBM: 20+ quantum systems in their cloud, contracts with JPMorgan, ExxonMobil, Daimler.[20] Their Heron processor[14] delivers 133 qubits with improved error performance.
- Google: Claimed quantum supremacy in 2019,[2] and demonstrated below-threshold error correction in 2024.[21]
- IonQ:
    - IonQ went public via SPAC in 2021. Their Hyundai partnership, announced in 2022, remained active through 2024, focusing on quantum chemistry applications for battery materials. In 2024, IonQ released their Forte system featuring 35 algorithmic qubits—a metric that accounts for native gate performance rather than raw qubit count. An Airbus partnership, announced in 2023, applied quantum optimisation to aircraft loading problems.

Quantum Annealing:

- D-Wave: Systems deployed at Lockheed Martin, Google, NASA, Los Alamos. D-Wave went public in 2022 via SPAC on NYSE under ticker QBTS. By 2024 their Advantage2 prototype[15] demonstrated improved performance over Advantage system.
- Volkswagen: Traffic optimization in Beijing and Lisbon
- Denso: Manufacturing optimization
- NEC: Combinatorial optimization for logistics

These aren't research projects anymore. They're commercial deployments where quantum—imperfect, noisy, limited quantum—provides enough value to justify multimillion-dollar investments.

Two paths diverged in a quantum wood. One took precise, logical steps—calculating every quantum gate, building toward universal computation, chasing the dream of fault-tolerant quantum computers that could run Shor's algorithm and break RSA encryption.

The other rolled down energy landscapes—letting physics find solutions through quantum tunneling, solving optimization problems today rather than waiting for perfection tomorrow.

Both use the same quantum weirdness. Both face the same decoherence demons. Both require dilution refrigerators colder than outer space.

The gate model is the future—ten years away, always ten years away, until suddenly it's here and changes everything.

Quantum annealing is the present—limited, specialized, but working right now for companies willing to pay for it.

In the end, we probably need both. Different tools for different problems. The chess player and the marble roller, both climbing their respective paths toward the same summit.

The race has no obvious front-runner. Gate-model systems are more versatile but further from practical utility. Quantum annealers are more limited but commercially deployed today. Hybrid approaches are closing the gap between theory and practice.

What's certain is this: the engineering challenges of the previous chapters—the dilution refrigerators, the decoherence, the fabrication yields—apply to both approaches equally. The physics doesn't care which architecture you prefer.

What matters is whether the quantum advantage justifies the cost for your specific problem.

That question has no general answer.

Only specific ones.

Turn the page.

References

1 Shor, P. W. (1997). Polynomial-time algorithms for prime factorization and discrete logarithms on a quantum computer. *SIAM Journal on Computing*, *26*(5), 1484–1509.

2 Arute, F., et al. (2019). Quantum supremacy using a programmable superconducting processor. *Nature*, *574*(7779), 505–510.

3 Nielsen, M. A., & Chuang, I. L. (2010). *Quantum computation and quantum information* (10th anniversary ed.). Cambridge University Press.

4 Farhi, E., Goldstone, J., Gutmann, S., Lapan, J., Lundgren, A., & Preda, D. (2001). Quantum computation by adiabatic evolution. arXiv:quant-ph/0001106.

5 Kadowaki, T., & Nishimori, H. (1998). Quantum annealing in the transverse Ising model. *Physical Review E*, *58*(5), 5355–5363.

6 D-Wave Systems. (2023). Annual report and corporate milestones. NYSE: QBTS Investor Relations.

7 Rønnow, T. F., Wang, Z., Job, J., Boixo, S., Isakov, S. V., Wecker, D., Martinis, J. M., Lidar, D. A., & Troyer, M. (2014). Defining and detecting quantum speedup. *Science*, *345*(6195), 420–424.

8 Albash, T., & Lidar, D. A. (2018). Adiabatic quantum computation. *Reviews of Modern Physics*, *90*(1), 015002.

9 Volkswagen Group. (2019). Traffic optimization using quantum computing [Press release].

[10] Denso Corporation. (2021). Quantum annealing for manufacturing optimization. Technical Report.

[11] Benedetti, M., Realpe-Gómez, J., Biswas, R., & Perdomo-Ortiz, A. (2016). Estimation of effective temperatures in quantum annealers for sampling applications: A case study with possible applications in deep learning. *Physical Review A, 94*(2), 022308.

[12] Applegate, D. L., Bixby, R. E., Chvátal, V., & Cook, W. J. (2006). *The traveling salesman problem: A computational study*. Princeton University Press.

[13] Grover, L. K. (1996). A fast quantum mechanical algorithm for database search. *Proceedings of the 28th Annual ACM Symposium on Theory of Computing*, 212–219.

[14] IBM Quantum. (2024). IBM Quantum Heron processor architecture. IBM Research Blog.

[15] D-Wave Systems. (2023). D-Wave Advantage system datasheet. D-Wave Technical Documentation.

[16] McGeoch, C. C., & Farre, P. (2021). The D-Wave Advantage system: An overview. D-Wave Systems White Paper.

[17] IBM Quantum. (2024). IBM Quantum development roadmap 2024. IBM Research.

[18] IBM Quantum. (2023). Qiskit Runtime: Quantum-classical iteration. IBM Research Documentation.

[19] Farhi, E., Goldstone, J., & Gutmann, S. (2014). A quantum approximate optimization algorithm. arXiv:1411.4028.

[20] IBM Quantum Network. (2024). Partners and collaborations. IBM Corporate.

[21] Google Quantum AI & Collaborators. (2024). Quantum error correction below the surface code threshold. *Nature, 614*(7949), 676–681.

[22] National Institute of Standards and Technology. (2024). Post-quantum cryptography: Selected algorithms. NIST Special Publication 800–208.

PART IV

QUANTUM AT WORK—ALGORITHMS, SECURITY & BEYOND

2024: Grover's algorithm running on IBM hardware.

2034: Shor's algorithm may threaten RSA-2048.

The decade between is borrowed time.

CHAPTER

11

Quantum Algorithms That Actually Work

Wrong answers cancel out. The right one remains

What's an Algorithm, Anyway?

An algorithm is just a recipe. Nothing more exotic than that. Your GPS uses one to find the fastest route home. Your bank uses one to catch the fraudulent transaction before you do. The spam filter reading your email uses one that's been trained on billions of messages to recognise the particular fingerprints of Nigerian princes in financial distress.

Every algorithm humans have built for the past eighty years shares one fundamental characteristic: it follows a single path through a problem. One instruction at a time. One possibility evaluated, then the next. The computer is fast—extraordinarily fast—but it's still marching through possibilities in a line, the way you'd search for your keys by checking one coat pocket, then another, then the kitchen counter, then retracing your steps.

Quantum algorithms break this rule in a way that still feels slightly unbelievable even to the people building them.

Instead of following one computational path, they explore many possibilities simultaneously through superposition. But here's where most explanations stop—and where the real magic begins.

Superposition alone doesn't give you anything useful. Put all your answers in superposition, measure randomly, and you get a random wrong answer. No better than guessing. The quantum computer hasn't solved your problem—it's just generated expensive noise.

> *The secret isn't superposition. It's what happens next. Quantum algorithms use interference—the same phenomenon that makes noise-canceling headphones work, that creates the patterns in oil slicks on wet pavement—to make wrong answers cancel each other out while the correct answer grows stronger.*

By the time you measure, the right answer isn't hiding among a million equally probable options. It's standing alone, amplified above everything else, practically announcing itself.

That shift—from searching to amplifying—is the entire secret of quantum computation. And it took physicists several decades to figure out how to control it deliberately enough to build algorithms from it.

The Thing Nobody Explains Properly

Here is where almost every quantum computing explanation—popular books, TED talks, university lectures—quietly misleads you.

The standard story goes like this: a quantum computer puts all possible answers in superposition simultaneously, checks them all at once, and out pops the right answer. It sounds like magic. It is also not quite right.

Think about what would actually happen if that were the whole story. You have a million possible answers, all in superposition. You measure. Quantum mechanics collapses that superposition to one random answer—any of the million, with equal probability. You would get the right answer one time in a million. Exactly as likely as guessing. Exactly as useless as a classical computer picking randomly.

> *Superposition is not the advantage. It is the setup. The advantage is what the algorithm does to the superposition before you measure.*

Imagine you are trying to find a specific person in a darkened stadium filled with a million people. One approach: shine a flashlight on each person individually until you find them. That is classical computing—exhaustive, sequential, and slow for large crowds.

A second approach: give everyone a flashlight and have them all shine at once. That is superposition—all possibilities illuminated simultaneously. But if everyone shines with equal brightness, you still cannot tell who you are looking for. The information is there but useless.

The quantum approach does something else entirely. You give everyone a flashlight, then use the physics of wave interference to make every flashlight except one cancel out. Wrong answers destructively interfere—their light waves arrive out of phase and annihilate each other. The correct answer constructively interferes—its waves arrive in phase and combine into a beam bright enough to see from across the stadium.

When you finally look, there is only one light left burning.

That is not magic. That is controlled wave mechanics applied to probability amplitudes.

Why the Oracle Matters

In quantum mechanics, every possible answer carries a probability amplitude—a number that determines how likely that answer is when you measure. When you first create the superposition, all amplitudes are equal. Every answer has the same tiny probability. Measure now and you get a random result. Useless.

The oracle—The oracle is the part of the algorithm that can recognise the correct answer when it encounters it—not by knowing it in advance, but by checking a condition. It does not tell you the answer directly. Instead it flips the sign of the correct answer's amplitude, marking it mathematically without revealing it.

Nothing useful has happened yet.

Then comes the diffusion step. The diffusion operator reflects all amplitudes around their average value. Because the correct answer's amplitude is now negative while everything else is positive, this reflection has an asymmetric effect. It pushes the correct answer's amplitude upward while pushing every wrong answer's amplitude downward.

The correct answer gets a little brighter. Everything else gets a little dimmer.

Repeat this oracle-and-diffusion cycle approximately √N times and the correct answer's amplitude has grown to nearly one while everything else has collapsed toward zero. Measure now and you get the right answer with overwhelming probability.

You did not check all the answers. You amplified the right one into visibility.

This mechanism—phase marking followed by amplitude amplification through interference—is the engine inside every useful quantum algorithm. Grover's uses it for search. Shor's uses the Quantum Fourier Transform to achieve the same effect for period finding. VQE uses it to converge on molecular ground states. QAOA uses it to push optimization landscapes toward low-energy solutions.

Once you see this pattern, quantum computing stops looking like mysterious parallel processing and starts looking like very precise wave engineering. The qubits are not checking all answers simultaneously and handing you the right one. They are being orchestrated so that wrong answers cancel themselves into silence while the right answer shouts.

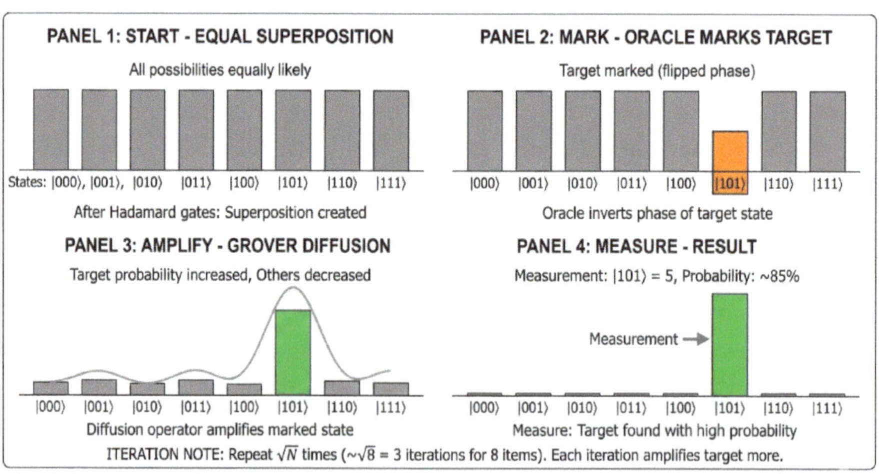

Grover's algorithm in 4 steps: (1) Create equal superposition of all states. (2) Oracle marks target by inverting its phase. (3) Grover diffusion amplifies marked state through interference. (4) Measure—target state has ~85% probability. Repeat ~$\sqrt{N}$ iterations for N items.

The Phone Book Problem

Here's the challenge: You're searching for a specific name in a phone book with one million entries.

A classical computer checks names one by one. On average, it'll check 500,000 entries before finding the right one. If it's unlucky and the name is last, all million.

A quantum computer using Grover's algorithm finds it in roughly 1,000 tries—the square root of the total.[1]

Not by magic. Not by knowing the answer in advance. By exploiting quantum interference to amplify the probability of the correct answer while suppressing all the wrong ones.

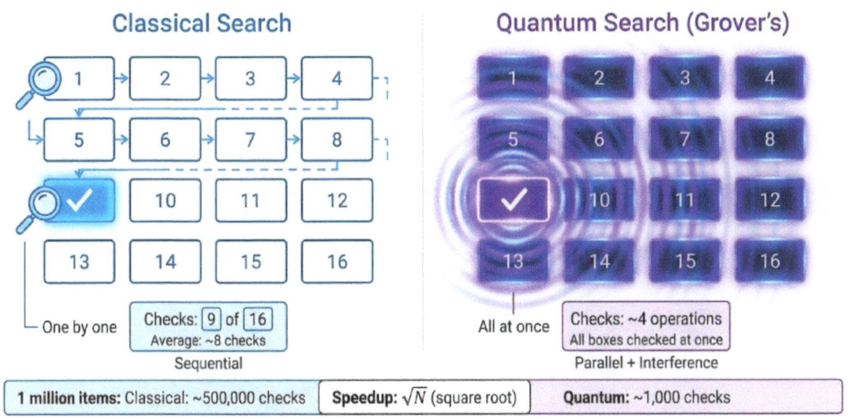

Classical search checks items one by one (average: N/2 checks). Grover's quantum search uses superposition to evaluate all items simultaneously, then interference to amplify the correct answer. Result: ~$\sqrt{N}$ checks. For 1 million items: 500,000 vs 1,000 checks.

Think of it like this: Instead of checking each grain of rice individually, you tap the bowl and listen for which grain vibrates out of phase with all the others. One grain's response is subtly different—not louder, but offset. The interference pattern that emerges when all grains respond together makes that one grain visible without checking any of them directly.

CLASSICAL VS QUANTUM SEARCH (1 MILLION ITEMS)		
Approach	**Search Time**	**Analogy**
Classical	500,000 ops (average)	Checking each grain of rice individually
Quantum (Grover's)	1,000 ops (√N)	Listening for the grain that *resonates differently*

Quantum computers don't magically "know" the answer—they amplify the probability of the right one through interference.

The Secret: Interference

In classical logic, each operation adds or multiplies numbers. Straightforward. Deterministic.

In quantum logic, each operation manipulates amplitudes—complex numbers that can add constructively or cancel destructively, like waves in water.

Think of waves in a pond:

- If two ripples meet in sync, they form a big wave (constructive interference).
- If they meet out of sync, they cancel (destructive interference).

Quantum algorithms choreograph this interference deliberately. They arrange the computation so wrong answers interfere destructively and vanish, while the right answer interferes constructively and grows stronger.

By the end of the calculation, when you measure, the correct answer has the highest probability. Not 100%—this is quantum mechanics, nothing is certain—but high enough that running the algorithm a few dozen times gives you the right answer with confidence.

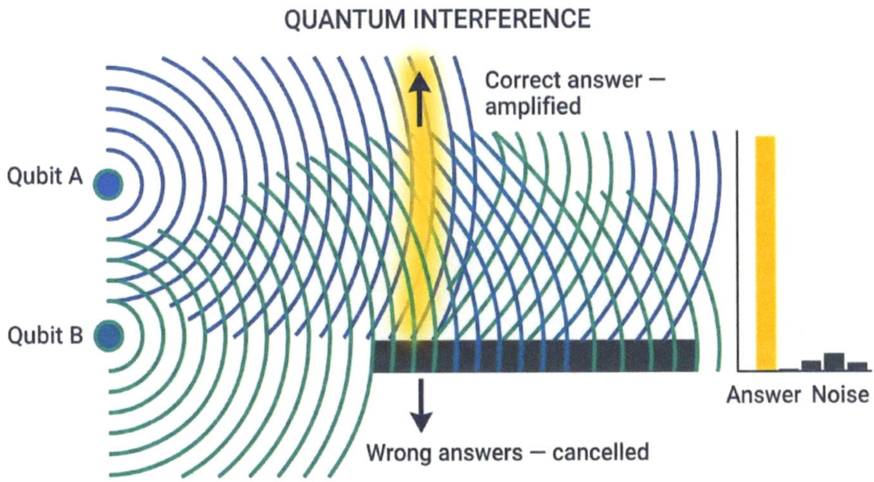

Classical computers try paths one at a time. Quantum computers arrange their calculation so that wrong paths cancel each other out before you ever look—and only the right path remains.

That's not faster computation. That's fundamentally different computation.

Quantitative Peek: Why It's So Powerful

Let's take a simple estimation.

For a classical computer searching a database of one trillion items (N = 10^{12}) at one billion operations per second:

System	Formula	Calculation	Time
Classical	$\text{Time} = \frac{N}{\text{speed}}$	$\frac{10^{12}}{10^9} = 10^3$	≈ 1,000 seconds (≈ 17 minutes)
Quantum (Grover's)	$\text{Time} = \frac{\sqrt{N}}{\text{speed}}$	$\frac{10^6}{10^9} = 10^{\ 3}$	≈ 0.001 seconds

Same problem. Same computer speed. One thousand seconds versus one millisecond. The difference isn't processing power—it's mathematical orchestration.

The Building Blocks: Quantum Gates

To make this happen, quantum programmers use quantum gates—the building blocks of quantum circuits. In classical computers, gates are like little switches:

- AND, OR, NOT—simple logic.

In quantum computers, gates are *rotations* that twist and turn qubits in their weird probability space.

For example:

- **Hadamard gate (H):** Puts a qubit into superposition—both 0 and 1.
- **CNOT gate:** Entangles two qubits so their fates are linked.
- **Pauli gates (X, Y, Z):** Flip, rotate, or spin the qubit like a dancer.

QUANTUM GATES—QUICK REFERENCE TABLE

Gate	Symbol	Function	Analogy
X	×	Flips qubit (like NOT gate)	Flipping a coin
H (Hadamard)	H	Creates superposition	Tossing coin in the air
CNOT	⊕	Entangles two qubits	Twins that copy each other's mood
T	T	Adds phase (quantum seasoning)	Adjusting taste
RX, RY, RZ	-	Rotations on Bloch sphere	Spinning the globe of probability

Grover's Algorithm: Quantum Search

In 1996, Lov Grover at Bell Labs developed an algorithm that searches unsorted databases with quadratic speedup.[2]

The algorithm works like this:

Step 1: Apply Hadamard gates to put all qubits into superposition. Every possible answer now has equal amplitude—equal probability. Nothing useful has happened yet.

Step 2: Apply an oracle that marks the correct answer by inverting its phase. You haven't measured anything—you've just tagged it mathematically.

Step 3: Apply Grover's diffusion operator, which amplifies the marked answer's probability while suppressing everything else through interference.

Step 4: Repeat steps 2–3 about √N times. Each iteration amplifies the correct answer more.

Step 5: Measure. The answer with highest probability is your result.

Run this a few dozen times to verify. The correct answer will dominate the statistics.

Grover's algorithm won't break RSA encryption. Grover weakens symmetric encryption by roughly a square-root speedup. It won't revolutionize most computing. But for specific search and optimization problems—database queries, cryptographic hash collisions, certain AI training loops—it provides genuine quadratic speedup.[3]

For one million items:

- Classical: ~500,000 checks on average
- Grover's: ~1,000 checks

For one billion items:

- Classical: ~500 million checks
- Grover's: ~31,623 checks

Not exponential. Quadratic. But quadratic across billion-element datasets is transformative.

Shor's Algorithm: Breaking the Internet

Now for the famous one. The one that made the NSA nervous.

In 1994, mathematician Peter Shor at Bell Labs developed a quantum algorithm that can factor large numbers into their prime components exponentially faster than any known classical algorithm.[4]

Why does this matter? Because RSA encryption—the cryptographic system protecting your bank account, email, medical records, and most of the internet—relies on the assumption that factoring large numbers is computationally infeasible.

Factoring the result back into those primes? Classical computers would take longer than the age of the universe for numbers used in real RSA keys.[5]

Shor's algorithm changes the game completely.

For a 2,048-bit RSA key (current internet standard):

- Classical factoring: Billions of years
- Shor's algorithm on a fault-tolerant quantum computer with ~4,000 logical qubits: Hours[6]

This isn't theoretical. It's mathematically proven to work. The only reason it hasn't broken RSA yet is that we don't have quantum computers powerful enough. We have dozens of qubits. We need thousands of error-corrected logical qubits.

But we're getting closer every year.

This is why governments and companies are frantically developing post-quantum cryptography—encryption algorithms that even quantum computers can't crack.[7] The NSA issued updated cryptographic guidance in 2022 requiring federal agencies to plan transitions to quantum-resistant encryption. NIST finalised the first post-quantum cryptographic standards in 2024.[8]

How Shor's Algorithm Actually Works

Shor's algorithm doesn't directly factor numbers. That's still hard even for quantum computers.

Instead, it transforms the factoring problem into a period-finding problem, which quantum computers can solve efficiently using the Quantum Fourier Transform.[9]

Here's the simplified version:

Step 1: Pick a random number and compute its powers mod N (the number you're trying to factor).

Step 2: Find the period—how often the pattern repeats. Classically, this is exponentially hard. Quantumly, using QFT, it's polynomial.

Step 3: Use the period to compute factors through classical number theory (Euclid's algorithm).

The quantum speedup happens in step 2. The Quantum Fourier Transform uses interference to amplify the signal at the correct period while suppressing all others—the same mechanism as Grover's, applied to frequency space rather than a database.

A classical Fast Fourier Transform requires operations that grow roughly as N log N—manageable but significant for large N. The Quantum Fourier Transform requires only $(\log N)^2$ operations.[10] That shift—from N log N to $(\log N)^2$—is what transforms billions of years into hours. Not faster hardware. Fundamentally different mathematics

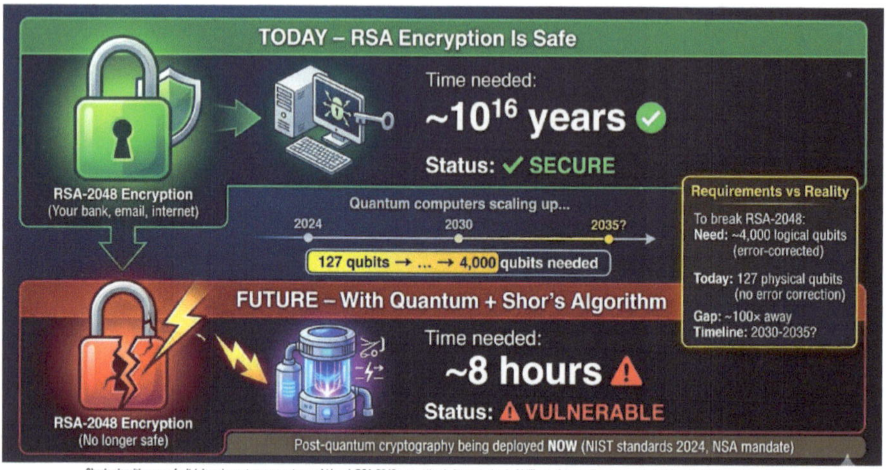

Shor's algorithm on a fault-tolerant quantum computer could break RSA-2048 encryption in hours instead of billions of years. Requirement: ~4,000 error-corrected logical qubits. Current systems: 127 physical qubits without error correction. Not immediate threat, but governments and companies are deploying post-quantum cryptography now in preparation.

VQE: Simulating Molecules

Not all quantum algorithms are about breaking things. Some create.

The Variational Quantum Eigensolver (VQE) simulates molecular structures to find their lowest-energy configurations—ground states.[11] This is essential for drug discovery, materials science, and chemistry.

Classical computers struggle with molecules beyond a dozen atoms. The quantum state space explodes exponentially. Simulating a 50-atom molecule requires tracking more computational states than there are stars in every galaxy astronomers have observed.

Quantum computers naturally represent quantum systems. A 50 qubits represent a space of 2^50 quantum states. No exponential blowup.

VQE is a hybrid algorithm: quantum computers prepare trial states, classical computers optimise parameters, repeat until convergence.[12]

In 2020, Google demonstrated VQE simulating diazene, a small nitrogen-hydrogen molecule, with chemical accuracy—meaning precise enough to be scientifically useful, not just approximately correct.[13] Two years later, IBM pushed this further with larger organic molecules relevant to battery chemistry.[14] Neither team is simulating complex pharmaceuticals yet.

This isn't sci-fi. Pharmaceutical companies are investing hundreds of millions in quantum chemistry simulations.[15] Not because it works perfectly today. Because it's the only path forward for complex molecular systems.

Why Variational Algorithms Work When Deep Circuits Fail

You have now seen VQE and QAOA described as working on today's hardware. In Chapter 12 we cover decoherence and error rates in full. But a preview is useful here: quantum computers lose coherence in microseconds and accumulate errors with every gate operation. So how do variational algorithms produce useful results at all?

The answer reveals something fundamental about how quantum computers are actually used today—and why the gap between what works now and what the field is ultimately aiming for is so large.

Recall that a quantum circuit with one thousand gates at 0.1% error rate per gate has roughly a sixty percent chance of producing garbage. For Shor's algorithm, which requires tens of millions of gates, the probability of getting a useful result is effectively zero on current hardware. The errors cascade until nothing meaningful remains.

Variational algorithms survive by never running a circuit that long[24].

The hybrid quantum-classical loop at the heart of VQE and QAOA works like this:

```
Step 1: Classical computer proposes initial circuit parameters
                            ↓
Step 2: Quantum computer runs a short, shallow circuit
        Measures energy or cost function value
                            ↓
Step 3: Classical computer receives measurement result
        Adjusts parameters using classical optimization
                            ↓
Step 4: Quantum computer runs again with new parameters
                            ↓
Step 5: Repeat 100 to 1,000 times
                            ↓
Step 6: Converge to optimal or near-optimal solution
```

Each individual quantum run is short—shallow enough that decoherence does not destroy the result before measurement. The classical optimiser handles the complexity that would otherwise require millions of quantum gates, iterating toward a solution across many brief quantum measurements rather than attempting one enormous quantum calculation.

Neither system alone could solve the problem. The quantum computer cannot run the deep circuit required for a direct solution. The classical computer cannot efficiently explore the quantum state space that gives the algorithm its advantage. Together, in this iterative partnership, they extract useful answers from hardware that should, by rights, be too noisy to trust.

This is why VQE works for small molecules today. The quantum circuit simply prepares a trial molecular state and measures its energy—a brief, manageable operation well within current coherence limits. The classical optimiser decides which trial state to prepare next. The loop converges to the molecular ground state through accumulated small measurements rather than one definitive quantum calculation.[12]

And this is precisely why Shor's algorithm cannot be adapted to run on today's hardware. The algorithm cannot be decomposed into a variational framework. It requires one unbroken quantum calculation—a single deep circuit with millions of gates—and no classical optimiser can substitute for that quantum computation in pieces. The calculation either runs to completion on a fault-tolerant quantum computer or it produces noise. There is no middle ground.

The timeline implications of this distinction are significant. Variational quantum algorithms—the class of algorithms suited to today's noisy, intermediate-scale quantum hardware—are available to researchers and companies today.[23] The results are limited in scale and noisy in practice, but they are real and improving year by year. Fault-tolerant algorithms like Shor's are at minimum a decade away from practical deployment, likely longer for the problem sizes that matter for cryptography.

Any conversation about quantum computing impact needs to specify which category. Near-term variational impact is already measurable in specific chemistry and optimization problems. Long-term fault-tolerant impact is real but distant. Conflating them—which almost every press release does—generates both false urgency and false disappointment depending on which audience receives the message.

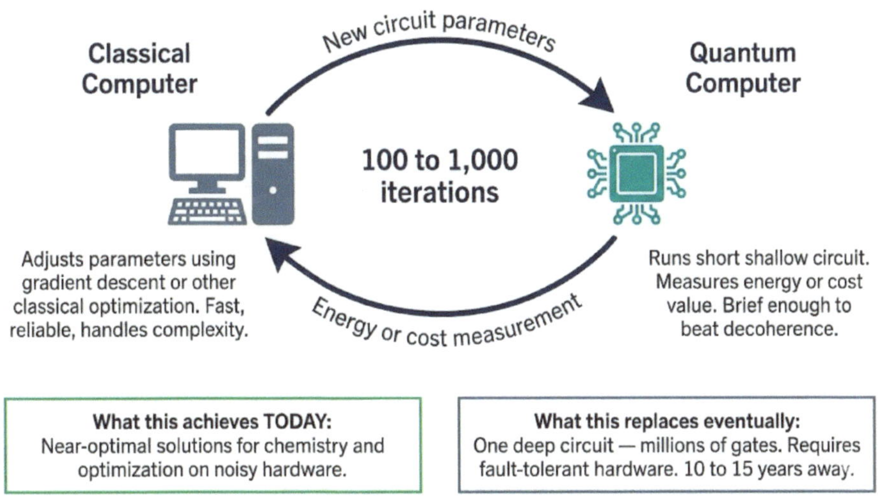

QAOA: Practical Optimization

The Quantum Approximate Optimization Algorithm (QAOA) bridges theory and practice more directly than any other quantum algorithm currently deployed.[16] Volkswagen used it to optimise traffic flow in Beijing and Lisbon.[17] Airbus applied it to aircraft loading.[18] Neither deployment achieved theoretical optimality. Both provided solutions good enough to justify the cost. For combinatorial optimization—problems with enormous solution spaces where near-optimal beats never-optimal—QAOA represents the most practically mature quantum algorithm available today.

Algorithm	Solves	Speedup	Real Use	Status
Grover's	Unstructured search	√N (quadratic)	Databases, AI search	Working on NISQ, limited scale
Shor's	Integer factorisation	Exponential	Cybersecurity (breaks RSA)	Requires fault-tolerant hardware
QFT (Quantum Fourier Transform)	Frequency/ periodicity detection	Exponential (for specific problems)	Signal processing, phase estimation	Component in other algorithms
VQE (Variational Quantum Eigensolver)	Molecular ground-state energy	Polynomial (quantum-advantage potential)	Chemistry, Pharma	Works for small molecules (NISQ era)
QAOA (Quantum Approximate Optimization Algorithm)	Combinatorial optimisation	Polynomial (heuristic)	Logistics, Finance	Commercial pilots and early deployments

Four Terms the Media Keeps Confusing

If you have followed quantum computing news over the past five years, you have encountered a cluster of similar-sounding claims arriving in quick succession. Quantum supremacy achieved. Quantum utility demonstrated. Quantum advantage still elusive. These are not synonyms. They represent four genuinely different milestones on a progression that the popular press routinely conflates, and the confusion leads to wildly distorted expectations about where quantum computing actually stands today.

Quantum Volume is a hardware metric developed by IBM.[22] It measures the largest square quantum circuit—equal width and depth—that a system can execute with better than fifty percent accuracy. Higher quantum volume means more capable hardware. Think of it as engine horsepower: a measure of what the machine can theoretically do, not what it has actually accomplished.

Quantum Supremacy, a term Google introduced in 2019—and one that some researchers have proposed replacing with 'quantum primacy' to avoid loaded connotations, though Google itself has not adopted the change. This refers to a quantum computer completing any task faster than the best available classical computer.[22] Google's Sycamore processor sampled a random quantum circuit in two hundred seconds that Google estimated would take ten thousand years on Summit, then the world's most powerful supercomputer. This was real. It was also a completely useless task—random circuit sampling has no practical application. Classical algorithms have since dramatically reduced the classical time estimate. The milestone was scientifically significant as proof that quantum hardware can outperform classical hardware at something. It was not, despite headlines, the moment quantum computing became useful.

Quantum Utility is IBM's 2023 term for something more modest and arguably more meaningful: quantum computations that produce accurate, scientifically useful results for problems genuinely difficult to simulate classically, even if the quantum system is not necessarily faster.[20] IBM's 127-qubit Eagle processor demonstrated this for certain condensed matter physics simulations. The results were real science. Whether they were faster than the best classical methods remains actively debated.

Quantum Advantage is the actual goal the entire field is working toward: a quantum computer solving a practically important problem both faster and more accurately than any classical computer can. As of 2025, this has not been achieved for any useful problem.

Milestone	Status	Meaning
Quantum Volume	Measured routinely	Hardware capability benchmark
Quantum Supremacy	Achieved (2019)	Quantum faster at a narrow task
Quantum Utility	Demonstrated (2023)	Useful results, not yet faster
Quantum Advantage	Not yet achieved	Useful and faster than classical

Most investors, journalists, and executives believe quantum advantage has already been achieved. It has not—for any practically useful problem. What exists is promising progress toward that goal, with demonstrable utility emerging along the way. Both statements are simultaneously true.

This distinction matters enormously before you read the next quantum computing press release, evaluate a vendor claim, or consider investing in the sector. The algorithms that would deliver genuine advantage—Shor's algorithm for cryptography, VQE for complex drug molecules at scale—require hardware that is still years or decades away from maturity. The variational algorithms delivering quantum utility today are real and valuable. They are not yet transformative.

Calibrating these expectations is not pessimism. It is the difference between making rational decisions about quantum technology and being sold something that does not yet exist.

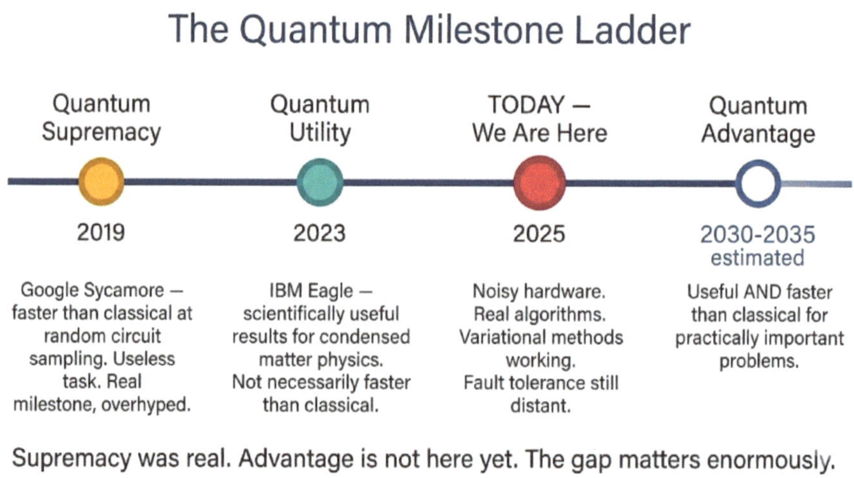

The Fine Print: What Quantum Doesn't Do

Quantum computers don't make everything faster.

They provide speedup only for problems with specific mathematical structure—periodicity for Shor's, unstructured search for Grover's, Hamiltonian simulation for VQE.

Quantum computers won't help with:

- Sorting lists (classical algorithms are already optimal)
- Running Excel spreadsheets
- Streaming Netflix
- Rendering graphics
- Most machine learning (classical GPUs are better)
- Anything requiring fast sequential operations

They're specialized accelerators for specific problem classes. Like GPUs revolutionized graphics and AI without replacing CPUs, quantum processors will revolutionize optimization, simulation, and cryptography without replacing classical computers.

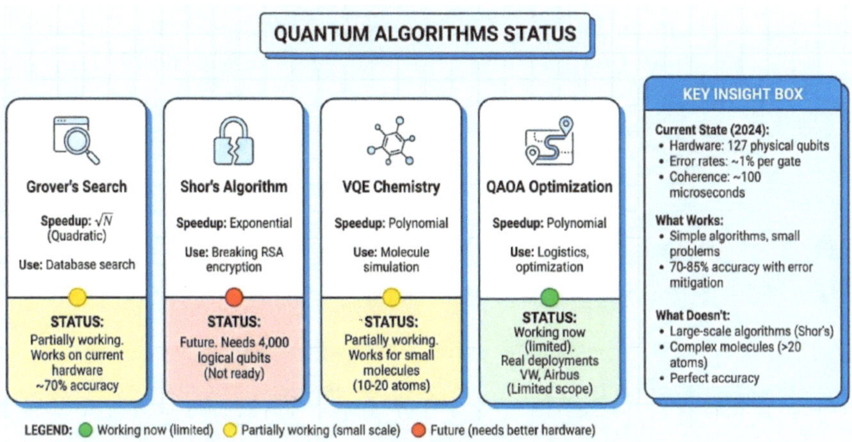

Quantum algorithm status (2024): Grover's and VQE work on current hardware with limited accuracy and scale. QAOA has real industrial deployments for specific optimization problems. Shor's requires fault-tolerant quantum computers not yet built (~4,000 logical qubits needed vs 127 physical qubits available).

For 99% of computing, your laptop is faster, cheaper, and better. For that crucial 1%—molecular simulation, certain optimization problems, post-quantum cryptography—quantum computers are the only option.

The distinction matters enormously for business decisions. Companies investing in quantum computing because it will make their databases faster or their spreadsheets more responsive are misallocating capital. Companies investing because they have specific molecular simulation, optimization, or cryptographic problems at scale are making rational bets. The difference is understanding not just what quantum computers can do, but what structure their speedup requires.

The Quantum Advantage Map

Telling you what quantum computers cannot do is only half the picture. The more useful skill—for researchers, investors, engineers, and anyone trying to evaluate quantum claims—is knowing how to recognise problems that do have the right mathematical structure. That recognition requires a framework, and it is one that almost no popular treatment of quantum computing provides.

Quantum speedup does not come from raw processing power. It comes from exploiting specific mathematical structures through interference. If your problem has those structures, quantum computing can provide genuine advantage. If it does not, quantum computing will not help regardless of how many qubits you have or how low your error rates are. The hardware is not the constraint for those problems—the mathematics is.

Problems Where Quantum Provides Real Advantage

Quantum Simulation is the most compelling near-term application and the area where quantum advantage is most theoretically certain. Simulating quantum systems—molecules, chemical reactions, material properties—requires tracking an exponentially large state space classically. A fifty-atom molecule has roughly two to the power of fifty quantum states to track simultaneously. No classical computer can store or manipulate that much information. A fifty-qubit quantum processor represents it directly, because both systems obey the same quantum mechanical laws. VQE is already producing useful results for small molecules, and the advantage grows exponentially with molecular complexity.

Factoring and Discrete Logarithms are where Shor's algorithm provides exponential speedup, mathematically proven and enormous in scale. The practical consequence—breaking RSA and elliptic curve cryptography—explains why NIST and the NSA have been urgently developing post-quantum standards since 2016. The speedup is real. The hardware to exploit it does not yet exist at the required scale.

Unstructured Search is Grover's territory. Quadratic speedup for finding items in databases with no organizing structure. Real and useful at scale, particularly for cryptographic applications requiring large search spaces. Not transformative in the way Shor's algorithm is, but genuinely valuable.

Certain Optimization Problems benefit from QAOA and quantum annealing. The advantage is heuristic and problem-dependent rather than universal—some combinatorial optimization problems respond well, others do not. For the right problems, the speedup is commercially significant, which is why Volkswagen and Airbus have made real deployments rather than keeping research purely academic.

Sampling from Complex Distributions is an area where quantum computers have a natural advantage because quantum states are inherently probability distributions. This has applications in machine learning, statistical physics, and financial modeling. Google's 2019 supremacy demonstration was technically a sampling problem, which is why the task itself was arcane even if the underlying capability was real.

Problems Where Quantum Provides No Advantage

Sequential Logic—any computation where each step depends on the result of the previous step—cannot benefit from quantum parallelism because the steps cannot be run in superposition. Most traditional algorithms fall here. Classical computers are already optimal.

Structured Data Retrieval—looking up a record by index, querying an organised database, accessing information you know the address of—is already O(1) classically. You cannot do better than instant. Grover's algorithm is for unstructured search, not organised retrieval.

General Machine Learning attracts enormous quantum hype but most neural network training involves matrix multiplications and gradient calculations that GPU clusters handle extremely efficiently. Quantum machine learning is a legitimate research area with specific potential applications, but quantum computers are not generally superior to classical GPU systems for training the kinds of models that power modern AI. The cases where quantum helps are narrow and specific.

Daily Computing—writing documents, streaming video, browsing the internet, running spreadsheets—does not involve computationally hard problems. It involves large volumes of simple operations that classical systems have been optimised for over seventy years. Quantum computers are strictly worse at all of it.

A Simple Test

Before anyone tells you quantum computing will transform their particular business, ask three questions:

Does the problem involve simulating quantum systems, factoring large numbers, or solving discrete logarithm problems? If yes, quantum advantage is mathematically proven and the question is only when hardware matures.

Does the problem involve unstructured search across enormous spaces, or sampling from complex probability distributions? If yes, quantum provides meaningful speedup for sufficient scale.

Does the problem involve sequential processing, organised database retrieval, standard machine learning training, or general-purpose computation? If yes, quantum will not help. The money is better spent on better classical hardware.

The companies making rational quantum investments are those with problems in the first two categories at scales large enough to matter. Everyone else is buying a solution for a problem quantum computing does not solve.

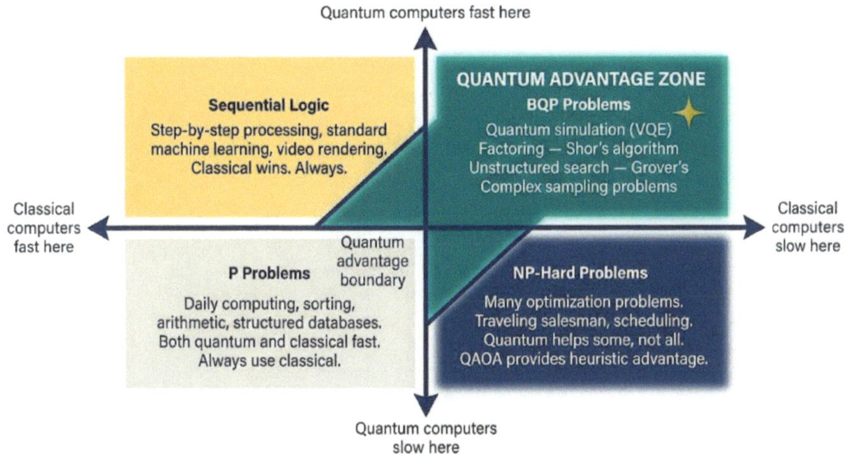

Quantum computers are not universally faster. They are specialized tools for a specific class of mathematical problems.

Writing Quantum Code (It's Easier Than You Think)

You can write quantum algorithms right now. IBM, Google, Amazon, and Microsoft all provide free cloud access to real quantum computers.[19]

Here's Grover's algorithm in Qiskit (Python):

```
from qiskit import QuantumCircuit
from qiskit.circuit.library import GroverOperator

# Create circuit for 3 qubits (8 possible states)
qc = QuantumCircuit(3, 3)

# Put qubits in superposition
qc.h([0, 1, 2])

# Apply Grover operator (marks and amplifies target)
oracle = create_oracle(target='101')  # Looking for |5⟩
qc.append(GroverOperator(oracle), [0, 1, 2])

# Measure
qc.measure([0, 1, 2], [0, 1, 2])
```

Twenty lines of Python. Runs on actual quantum hardware. Demonstrates quantum speedup for search. The code is accessible. The concepts are accessible. The barrier isn't understanding—it's hardware limitations.

Real-World Performance (The Honest Version)

Current quantum computers are noisy. Very noisy.

Grover's algorithm on IBM's hardware with 127 qubits? Works. Kind of. With ~70% accuracy after error mitigation.[20]

Shor's algorithm factoring RSA-2048? Requires ~4,000 error-corrected logical qubits. We have 127 physical qubits with no error correction. We're not even close.[6]

VQE simulating complex molecules? Works for small molecules (10–20 atoms). Larger systems accumulate too much error. QAOA optimizing delivery routes? Finds good solutions, often better than random guessing, sometimes worse than classical algorithms depending on the problem.[21]

The algorithms are proven. The math works. The hardware isn't ready yet. But "not ready yet" is different from "will never work."

Ten years ago, quantum computers had 5 qubits and couldn't run any algorithms. Today, 127 qubits running real algorithms with measurable (if limited) advantage. The trend is clear. The hardware is catching up. Slowly. Noisily. But catching up.

Quantum algorithms are strategies for harnessing nature's parallelism. They don't make computers faster at everything—they make them smarter at exploring possibilities. Classical computers march through solutions one at a time. Quantum computers dance through superposition, guided by interference, collapsing to answers with mathematical elegance.

We've spent 80 years building computers that think like we do—sequentially, deterministically, one step at a time. Quantum algorithms think like the universe does—in superposition, through interference, embracing uncertainty. That's not just a technological shift. It's a conceptual one.

Classical computers calculate. Quantum computers orchestrate. The algorithms exist. The mathematics is proven. Grover's is running on real hardware today—imperfectly, noisily, but running. Shor's is waiting for hardware that does not yet exist but gets measurably closer each year. VQE is producing chemistry results that matter. QAOA is optimizing real logistics for real companies paying real money for the privilege.

The gap between mathematically proven and practically deployed at scale is closing. Not as fast as press releases suggest. Faster than the skeptics predicted.

But when those answers are right, they're solving problems no classical computer ever could.

Turn the page.

References

[1] Grover, L. K. (1996). A fast quantum mechanical algorithm for database search. *Proceedings of the 28th Annual ACM Symposium on Theory of Computing,* 212–219.

[2] Grover, L. K. (1997). Quantum mechanics helps in searching for a needle in a haystack. *Physical Review Letters,* 79(2), 325–328.

[3] Boyer, M., Brassard, G., Høyer, P., & Tapp, A. (1998). Tight bounds on quantum searching. *Fortschritte der Physik,* 46(4–5), 493–505.

[4] Shor, P. W. (1994). Algorithms for quantum computation: discrete logarithms and factoring. *Proceedings of the 35th Annual Symposium on Foundations of Computer Science,* 124–134.

[5] Lenstra, A. K., et al. (2001). Selecting cryptographic key sizes. *Journal of Cryptology,* 14(4), 255–293.

[6] Gidney, C., & Ekerå, M. (2021). How to factor 2048-bit RSA integers in 8 hours using 20 million noisy qubits. *Quantum,* 5, 433.

[7] National Institute of Standards and Technology (2024). Post-quantum cryptography: selected algorithms. *NIST Special Publication,* 800–208.

[8] National Security Agency (2024). Quantum computing and post-quantum cryptography. NSA Cybersecurity Advisory.

[9] Shor, P. W. (1997). Polynomial-time algorithms for prime factorization and discrete logarithms on a quantum computer. *SIAM Journal on Computing,* 26(5), 1484–1509.

[10] Nielsen, M. A., & Chuang, I. L. (2010). *Quantum Computation and Quantum Information* (10th Anniversary Edition). Cambridge University Press.

[11] Peruzzo, A., et al. (2014). A variational eigenvalue solver on a photonic quantum processor. *Nature Communications,* 5, 4213.

[12] Cerezo, M., et al. (2021). Variational quantum algorithms. *Nature Reviews Physics,* 3, 625–644.

[13] Google Quantum AI (2020). Hartree-Fock on a superconducting qubit quantum computer. *Science,* 369(6507), 1084–1089.

[14] IBM Quantum (2023). Quantum chemistry simulations on IBM Quantum systems. IBM Research Technical Report.

[15] McKinsey & Company (2023). Quantum technology: sensing, computing, and communications. McKinsey Global Institute Industry Report.

[16] Farhi, E., Goldstone, J., & Gutmann, S. (2014). A quantum approximate optimization algorithm. arXiv:1411.4028.

[17] Volkswagen Group (2019). Quantum computing for traffic flow optimization in Beijing. Corporate Press Release.

[18] Airbus (2021). Quantum computing applications in aerospace loading optimization. Airbus Technical Report.

[19] IBM Quantum Experience, Google Quantum AI, Amazon Braket, Microsoft Azure Quantum (2024). Public cloud quantum access programmes. Company Technical Documentation.

[20] Kim, Y., et al. (2023). Evidence for the utility of quantum computing before fault tolerance. *Nature,* 618, 500–505.

[21] Harrigan, M. P., et al. (2021). Quantum approximate optimization of non-planar graph problems on a planar superconducting processor. *Nature Physics,* 17, 332–336.

[22] Cross, A. W., et al. (2019). Validating quantum computers using randomized model circuits. *Physical Review A,* 100, 032328.

[23] Arute, F., et al. (2019). Quantum supremacy using a programmable superconducting processor. *Nature,* 574, 505–510.

[24] Bharti, K., et al. (2022). Noisy intermediate-scale quantum algorithms. *Reviews of Modern Physics,* 94, 015004.

CHAPTER

12

Keeping Qubits Alive Long Enough to Actually Compute Something

Why quantum computers need 1,000 babysitter qubits for every one that does actual work

The Drama of Qubits

> *Quantum computers make errors constantly. Not occasionally, not in edge cases—constantly.*

Every gate operation has a small probability of failure. Every qubit is perpetually threatened by thermal fluctuations, electromagnetic noise, and cosmic radiation. Even in the most carefully engineered systems operating at temperatures colder than deep space, errors accumulate at rates that would make any classical computer engineer wince.

For a quantum computer running an algorithm that requires one thousand gate operations, with each gate failing just one percent of the time, the probability of the entire calculation completing without a single error drops below one in twenty-three thousand.

That's not a prototype problem or an engineering limitation that more funding will solve. It's physics. Quantum states are inherently fragile. Decoherence is inevitable.

So how do we build reliable quantum computers from components that self-destruct in microseconds?

The answer took three decades to develop, requires a thousand physical qubits for every one useful qubit, and is the single most important unsolved engineering challenge in quantum computing.

It's called quantum error correction.

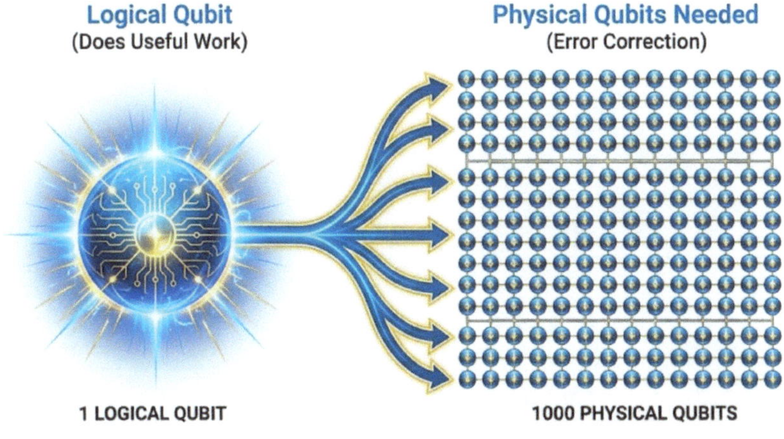

A single fault-tolerant logical qubit, capable of reliable computation, requires a massive array of physical qubits to perform continuous error correction and protect quantum information from decoherence.

How Fast Can a Qubit Die?

Here's how long different qubit types maintain quantum coherence before decoherence destroys them:

Superconducting qubits: ~100 microseconds (0.0001 seconds)[2]

This fragility has a name: Decoherence—the collapse of quantum states when the outside world intrudes—was covered in Chapter 6. What matters here is its rate: how many microseconds you have before it destroys your calculation.[1]

That's how long your coffee stays warm if you blink very slowly.

Trapped ion qubits: ~10 seconds[3]

Better! But still shorter than most YouTube ads.

Photonic qubits: ~1 microsecond[4]

Basically a camera flash.

Compare this to classical computer memory (RAM), which holds data for hours or days without power, or hard drives that store information for decades.

The Problem: You Can't Copy Quantum States

In classical computers, error correction is straightforward. Store three copies of each bit. If one flips from 0 to 1 due to cosmic radiation or electrical noise, the other two outvote it.

Bit copies: 0, 0, 0 → One flips → 0, 1, 0 → Majority vote → Correct answer: 0

Simple. Effective. Used in every classical computer, hard drive, and communication system on Earth.

Quantum computers can't do this.

Why? The No-Cloning Theorem, proven by Wootters and Zurek in 1982.[5] You cannot make a perfect copy of an unknown quantum state—not because of engineering limitations, but because the mathematics forbids it as surely as the speed of light forbids faster-than-light travel.

> *It's like trying to photocopy a dream: the moment you attempt it, the dream vanishes.*

Any solution to quantum error correction must work around this theorem, not through it. For a decade after it was published, many physicists believed it made practical quantum computing impossible.

They were wrong.

But the solution required rethinking what error correction actually means.

The Clever Workaround: Encoding

In 1995, Peter Shor (yes, that Peter Shor, the RSAbreaking algorithm guy) had a brilliant insight: you don't need to *copy* quantum information. You can *spread* it.[6]

Instead of duplicating one qubit into three identical copies, you encode its quantum information across multiple qubits in a carefully choreographed superposition.

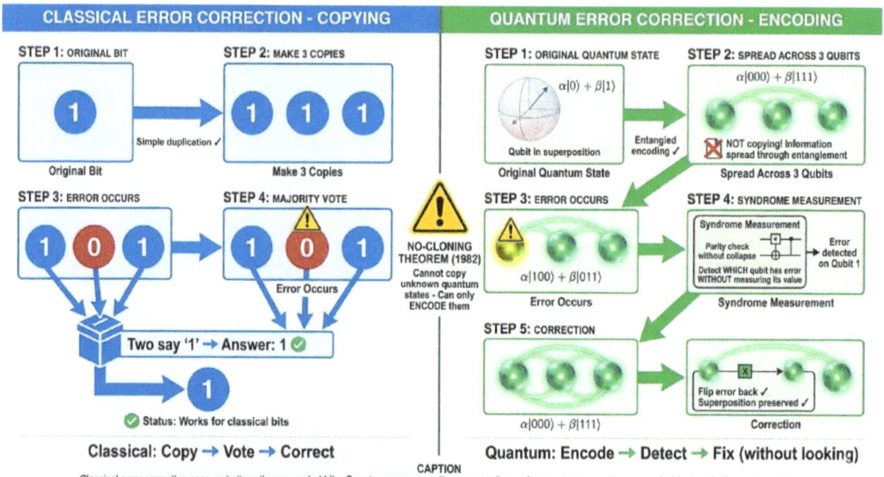

CAPTION
Classical error correction uses majority voting on copied bits. Quantum error correction uses parity syndrome measurements on encoded (not copied) quantum states. No-cloning theorem (1982) forbids copying quantum states, but allows spreading information across entangled qubits while preserving superposition.

Think of it like music. You can't duplicate a single singer's voice. But you can have a choir sing the same song. If one singer forgets a note or goes off-key, the harmony of the others preserves the melody. You don't need perfect individual singers—you need the collective to maintain the pattern.

This is quantum error correction. You don't save individual qubits. You save the *pattern* spread across many qubits.

The Simplest Example: Three-Qubit Bit-Flip Code

Here's the easiest quantum error correction code to understand.

You want to protect a single logical qubit in state $|\psi\rangle = \alpha|0\rangle + \beta|1\rangle$.

Encode it across three physical qubits:
$|\psi\rangle \rightarrow \alpha|000\rangle + \beta|111\rangle$

Notice: The superposition is preserved. The quantum information isn't copied—it's spread. If α and β represented complex probability amplitudes before, they still do. But now they're entangled across three qubits instead of one.

Now suppose one qubit flips due to noise:
$\alpha|000\rangle + \beta|111\rangle \rightarrow \alpha|100\rangle + \beta|011\rangle$

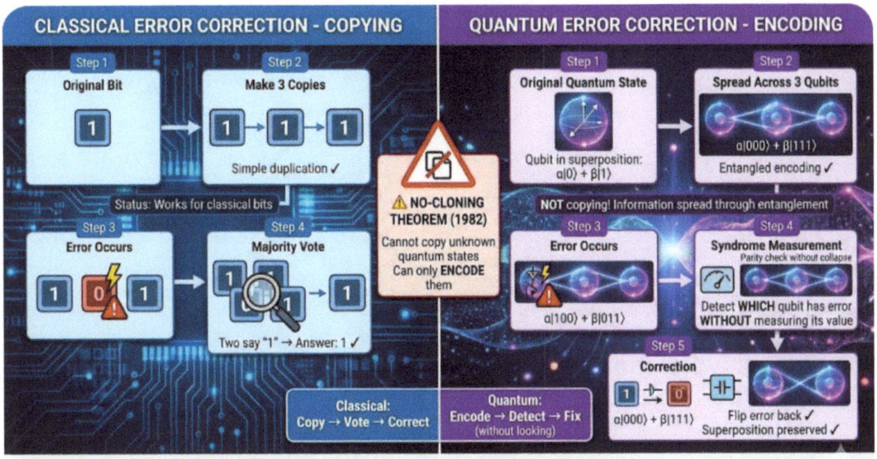

Classical error correction uses majority voting on copied bits. Quantum error correction uses parity syndrome measurements on encoded (not copied) quantum states. No-cloning theorem (1982) forbids copying quantum states, but allows spreading information across entangled qubits while preserving superposition.

You measure parity (checking if qubits agree) *without* measuring their actual values (which would collapse superposition). The parity measurements tell you *which* qubit flipped, not *what* it flipped to.

Then you flip it back. Error corrected. Superposition preserved.[7]

It's like three friends whispering the same secret. One mishears it. The other two compare notes—not revealing the secret itself, just checking consistency—and correct the one who got it wrong before rumors spread.

The Three Types of Quantum Errors

The three-qubit code above protects against one specific type of error: bit flips. But quantum computers face three distinct error types, and understanding all three explains why error correction is so much more expensive in quantum systems than in classical ones.

Bit-flip errors are the quantum equivalent of classical memory errors. A qubit in state $|0\rangle$ accidentally becomes $|1\rangle$, or vice versa. Caused by environmental noise, cosmic rays, or vibrations. This is the error type classical computers also face, and the three-qubit code above handles it.

Phase-flip errors have no classical equivalent. Here the qubit does not change between zero and one—it remains in superposition—but the relationship between its zero and one components shifts sign. A state that was $|+\rangle = (|0\rangle + |1\rangle)/\sqrt{2}$ becomes $|-\rangle = (|0\rangle - |1\rangle)/\sqrt{2}$. The probabilities of measuring zero or one are identical in both cases. You cannot detect a phase flip by simple measurement. But the phase relationship is precisely what interference algorithms depend on. A phase-flipped qubit will interfere incorrectly, producing wrong answers that look superficially like right answers.

This is why quantum error correction is so expensive. Classical computers only need to protect against bit flips. Quantum computers need to protect against bit flips and phase flips simultaneously, using separate correction codes for each. Shor's nine-qubit code—which we will see momentarily—handles both. But it requires nine physical qubits to protect one logical qubit, compared to three for classical majority voting.

Shor's nine-qubit code was the first to handle both error types simultaneously.[6] Steane independently developed a more efficient seven-qubit alternative the following year.[8]

Depolarizing errors combine both. The qubit's state becomes completely randomized—simultaneously experiencing bit flip, phase flip, or both with equal probability. This is the worst-case scenario, equivalent to the qubit forgetting everything it was doing and picking a random state instead.

The critical insight is this: phase errors destroy interference. Since interference is the source of quantum speedup—as Chapter 11 established—phase errors do not merely corrupt data in the way classical errors do. They eliminate the quantum advantage entirely. A quantum computer with uncorrected phase errors is not a slow quantum computer. It is a classical computer with extra steps.

This is the deeper reason quantum error correction requires such enormous overhead. You are not just protecting bits. You are protecting the wave-mechanical relationships between bits that make quantum computation possible in the first place.

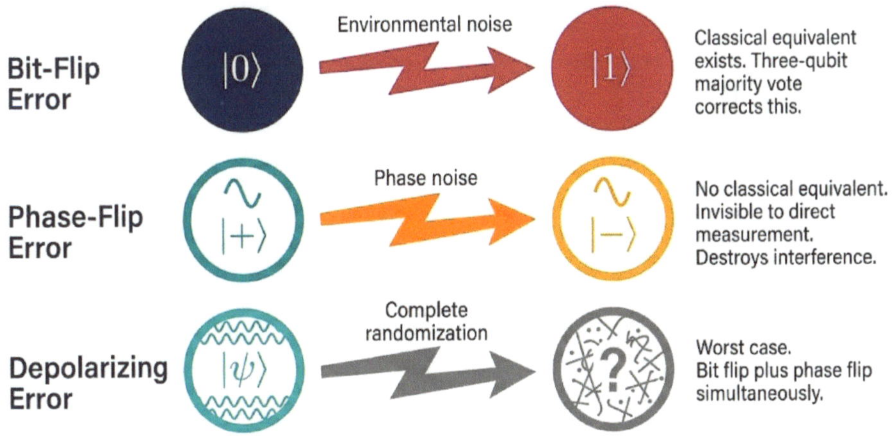

Classical computers face Row 1 only. Quantum computers face all three.
This is why quantum error correction needs 1,000 qubits to protect one.

The Real Challenge: Phase Errors

Bit flips ($|0\rangle \leftrightarrow |1\rangle$) aren't the only errors. Quantum states also suffer phase errors—where the relative phase between $|0\rangle$ and $|1\rangle$ components gets scrambled without the bit itself flipping.

A phase error is subtler than a bit flip. The qubit hasn't changed between zero and one—it's still in superposition. But the relationship between the zero and one components has shifted, like a musical chord where the notes are correct but the timing between them is off. The melody sounds wrong even though every individual note is right.

You need separate error correction for phase errors. Shor's nine-qubit code protects against both bit flips and phase flips.[6] One logical qubit requires nine physical qubits—and constant syndrome measurements to detect and correct errors continuously.

Nine qubits to protect one. That's overhead. And it gets worse.

Surface Codes: The *Industrial Standard*

For real quantum computers, the leading error correction approach is the surface code.[9]

Imagine a checkerboard of qubits. Data qubits sit on black squares. Measurement qubits sit on white squares. The measurement qubits constantly check their neighbouring data qubits—not reading their values, just detecting "weirdness" (parity violations).

If a data qubit has an error, nearby measurement qubits flag it. Classical computers decode these syndrome measurements in real-time and apply corrections—flipping qubits back or adjusting phases—before errors cascade.

It's like a neighbourhood watch for quantum information. Everyone's watching everyone, gossiping about suspicious activity, but nobody reveals actual secrets. The system self-heals through collective surveillance.

Surface codes are scalable, well-understood, and compatible with current hardware layouts. Google, IBM, and most major quantum computing efforts use surface codes or variants.[10]

But here's the brutal part: protecting one logical qubit with surface codes requires roughly 1,000 physical qubits.[9]

The elegance of surface codes is that they're local. Each measurement qubit only needs to interact with its immediate neighbours on the checkerboard, not with every qubit in the system. This locality makes surface codes compatible with physical hardware, where qubits can only directly interact with adjacent qubits.

The threshold theorem underlies all of this. In 1996, mathematicians proved that if physical gate error rates stay below a certain threshold—approximately one percent for surface codes—then adding more physical qubits will always reduce logical error rates.[24] Below threshold, scaling up works. Above threshold, scaling up makes things worse. For most of the past decade, quantum hardware sat above or right at the threshold, meaning error correction was theoretically possible but practically counterproductive. Google's Willow chip, discussed at the end of this chapter, was the first demonstration that a real quantum processor had crossed below threshold at scale. It is why December 2024 was a genuine milestone rather than another press release.

The Overhead Problem

Current quantum computers:

System	Phys. Qubits	Key Feature
IBM Condor	1,121	Largest by qubit count
Google Willow	105	Error-correction optimised
IonQ Forte	32	Trapped-ion architecture

To run Shor's algorithm and factor RSA-2048 encryption requires ~4,000 *logical* qubits.[14]

With 1,000:1 overhead from surface codes, that's 4,000,000 physical qubits.

We have hundreds. We need millions.

That's why practical quantum computing is still years away. The algorithms work. The error correction theory is proven. The hardware just isn't there yet.

Jay Gambetta, IBM Fellow and Vice President of Quantum Computing, has written extensively about the hardware requirements for fault-tolerant quantum computing. In a 2023 IBM Research blog post, he outlined the gap between current systems and fault tolerance: current IBM processors achieve 99.5% two-qubit gate fidelity in ideal single-gate conditions. Sustaining that fidelity across millions of simultaneous operations at the scale[15] required for error correction—that is the unsolved problem.

The numbers tell the story. IBM's largest current system has 1,121 physical qubits. Running Shor's algorithm against RSA-2048 encryption requires approximately four million physical qubits in a surface code implementation.[15] That's a gap of four orders of magnitude—not a refinement problem but a fundamental scaling challenge requiring manufacturing techniques that don't yet exist.

That gap—hundreds to millions, 99.5% to 99.99%—is everything.

The Path to Fault-Tolerant Quantum Computing

Understanding the gap requires seeing the full architecture of what fault-tolerant quantum computing actually looks like—not just the qubit counts but the layered structure that translates noisy physical hardware into reliable logical computation.

Think of it as four floors of a building, each converting something imprecise into something more reliable:

Floor 1—Physical Qubits: Millions of noisy, error-prone physical qubits. Gate fidelity around 99.9%. Coherence measured in microseconds. These are what IBM, Google, and IonQ are building today.

Floor 2—Logical Qubits: Each logical qubit encoded across roughly one thousand physical qubits using surface codes. Gate fidelity increases to approximately 99.9999%—six nines instead of three. Errors are continuously detected and corrected. The logical qubit can in principle operate indefinitely as long as the physical layer below it keeps functioning.

Floor 3—Fault-Tolerant Gates: Logic operations performed directly on logical qubits, where the gates themselves are error-corrected. This is harder than it sounds—performing a gate on an encoded qubit requires carefully orchestrated operations across all one thousand physical qubits simultaneously without accidentally introducing new errors.

Floor 4—Algorithms: Shor's algorithm, Grover's search, VQE at scale—running on logical qubits with error rates low enough that million-gate circuits produce meaningful answers.

Today we are building Floor 1 and learning to construct isolated examples of Floor 2. Floors 3 and 4 are engineered designs waiting for hardware to mature. The distance between demonstrating one logical qubit—which Google's Willow achieved—and running Shor's algorithm on four thousand logical qubits is the distance between striking one note cleanly on a piano and performing a symphony.

Both require the same instrument. The gap between them is entirely a matter of scale.

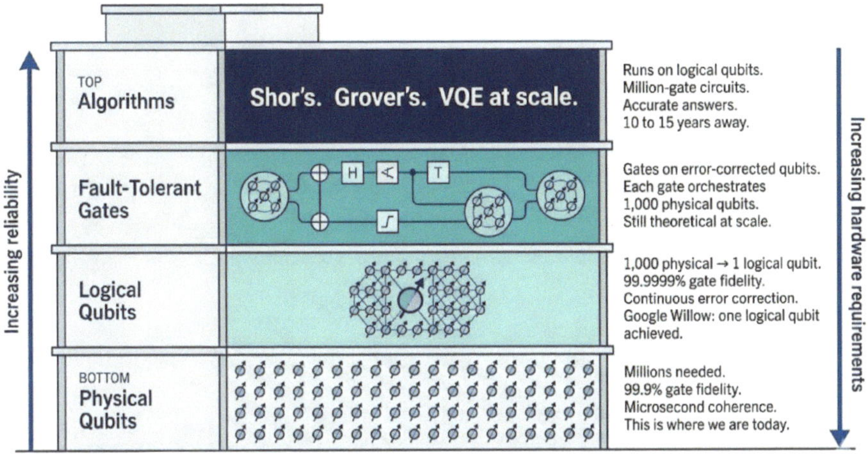

We are building Floor 1. Floor 4 is the goal. The distance between them is everything.

The Math: Why Errors Compound

Single-qubit gate error rate: 0.1% (0.001)
Two-qubit gate error rate: 1% (0.01)

For a quantum circuit with 1,000 gates (typical for useful algorithms):

Without error correction, for a 1,000-gate circuit at 0.1% error rate per gate:

Total error $\approx 1 - (0.999)^{1000} \approx 63\%$

Your algorithm has a sixty-three percent chance of producing garbage. For practical purposes, useless.

Most useful quantum algorithms require far more than 1,000 gates. Shor's algorithm for RSA-2048 requires approximately ten million gates. At 0.1% error rate per gate, the probability of getting a correct answer without error correction is effectively zero. Not low—zero. The algorithm is mathematically beautiful and practically useless without error correction at that scale.

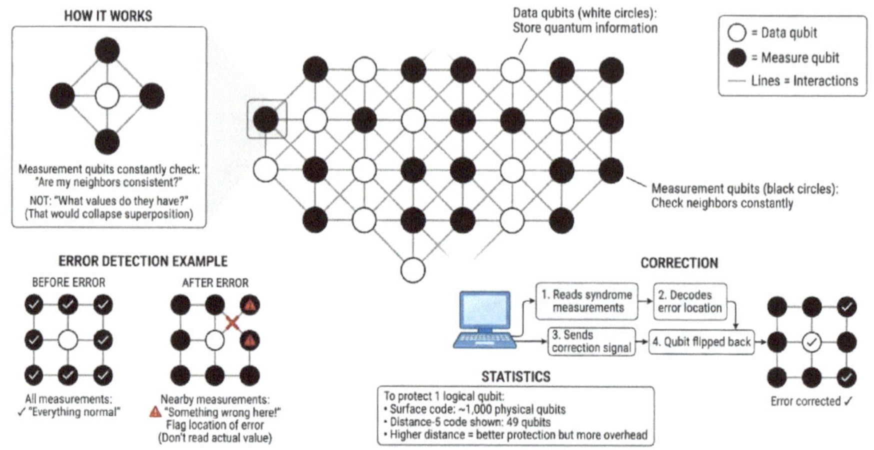

Surface code error correction: Data qubits (○) store quantum information. Measurement qubits (●) constantly check neighboring data qubits for parity violations without reading their values (which would collapse superposition). When error detected, classical computer decodes syndrome and corrects error. Overhead: ~1,000 physical qubits to protect 1 logical qubit.

With surface code error correction (1,000:1 overhead):

Logical gate error rate: ~10^{-6} (0.0001%)

Total error ≈ $1-(0.999999)^{1000} \approx 0.1\%$

Now it works. That is the difference between a phone call through static and a face-to-face conversation. Same words. One of them is actually usable.

But it costs 1,000× more qubits. That's the trade: perfection for resources.[23]

Error Mitigation: Band-Aids for Now

Full quantum error correction requires millions of qubits we don't have. So researchers developed error *mitigation*—techniques that improve results from noisy quantum computers without full error correction.[17]

Zero-noise extrapolation: Deliberately add extra noise to your quantum computation. Run it at multiple noise levels. Extrapolate what the result would be at zero noise mathematically.[18] It's like tasting soup at three different spice levels and guessing the perfect seasoning.

Measurement error mitigation: Characterize your measurement errors statistically, then correct the classical data after measurement.[19] Like adjusting your bathroom scale because you know it reads 2 pounds heavy.

Dynamical decoupling: Insert extra pulses between gates to counteract environmental noise, like noise-canceling headphones for qubits.[20]

These techniques don't prevent errors. They work around them. But they improve accuracy from 70% to 90%+ for current algorithms—enough to get useful results from noisy hardware.[21]

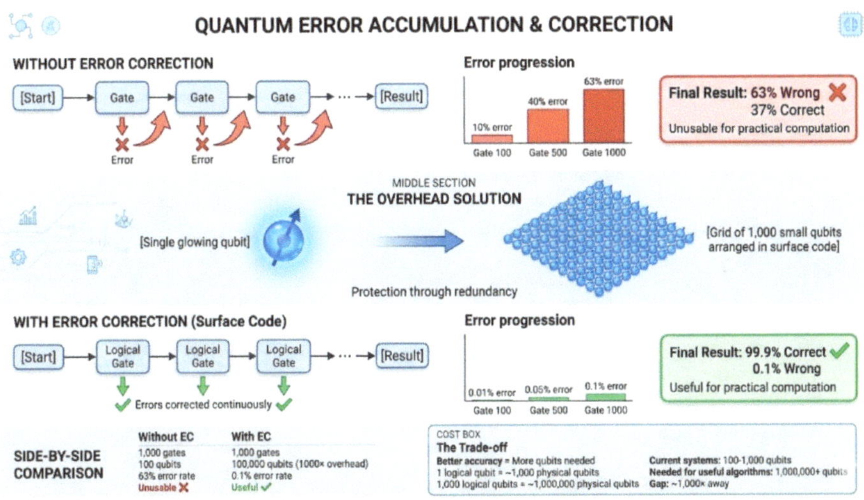

It's like image enhancement. Doesn't remove the noise entirely, but makes the signal clear enough to extract useful results.

Real Example: VQE on Noisy Hardware

In 2017, IBM researchers ran the Variational Quantum Eigensolver (VQE) algorithm to simulate a hydrogen molecule on a superconducting quantum processor.[22]

Without error mitigation:

- Accuracy: ~78%
- Chemical accuracy: No

With error mitigation (measurement error correction + extrapolation):

- Accuracy: ~94%
- Chemical accuracy: Yes (within 1 kcal/mol)

Same hardware. Better math. Useful chemistry results from a noisy quantum computer.

That's why error mitigation matters now, even before we achieve fault tolerance. It's the bridge between noisy prototypes and practical applications.

Classical Computers Babysitting Quantum Ones

Here's the irony: quantum computers need classical computers to survive.

Classical systems monitor qubits in real-time, decode syndrome measurements (which qubits have errors), compute corrections (which operations to apply), and execute feedback—all within microseconds, before errors cascade.[24]

The feedback loop must happen fast. Syndrome measurement, classical decoding, correction calculation, and physical qubit correction must all complete within the qubit's coherence time—typically microseconds for superconducting systems. This requires classical processors running error correction decoders at the edge of the quantum system, physically close to the dilution refrigerator, operating faster than any conventional computing architecture was designed for.

It is one of the more underappreciated engineering challenges in quantum computing: the classical control system required to keep quantum computation alive is itself a frontier engineering problem.

Google's Breakthrough: Below Threshold

In December 2024, Google's Willow chip demonstrated below-threshold error correction for the first time at scale—proof that adding physical qubits reduces logical error rates[12] rather than compounding them. The details are in Chapter 6.

Previously, adding more qubits often added more noise, making things worse. Google showed that with good enough hardware (99.7% two-qubit gate fidelity) and surface codes, scaling up reduces errors as theory predicts.

What it means here: the threshold theorem works in practice, not just theory. One logical qubit achieved. Thousands needed. But it's also proof of how hard this is: Google's best chip, with years of optimization, achieved this with 105 qubits creating one logical qubit. One.

To run useful algorithms requires thousands of logical qubits. We're celebrating getting one to work properly.

The road ahead is long.

Quantum error correction is humanity's most optimistic engineering challenge.

We're building computers from components that self-destruct in microseconds, correcting their errors using systems that require 1,000× overhead, running algorithms that need millions of qubits we haven't built yet—all to solve problems that classical computers fundamentally cannot.

It's absurd. Audacious. Probably necessary.

Because the universe operates on quantum mechanics. Molecules are quantum. Chemistry is quantum. Materials, photosynthesis, superconductivity—all quantum. Classical computers can't simulate quantum systems beyond a few dozen particles. The exponential complexity becomes insurmountable.

If we want to design better batteries, more efficient solar cells, room-temperature superconductors, or understand protein folding for drug discovery, we need quantum computers. And for quantum computers to work, we need error correction.

So we're building dilution refrigerators colder than space, fabricating qubit chips in rooms cleaner than operating theaters, writing error correction codes that spread information across thousands of entangled qubits, and running classical computers to babysit the quantum ones.

Qubits will always be fragile and decoherence will always threaten. But with enough cleverness, redundancy, and the particular kind of stubborn determination that has characterized every hard engineering problem humanity has eventually solved, we can extract signal from noise, truth from chaos, and reliable computation from the most delicate physical systems ever deliberately constructed.

It's not about perfection. It's about graceful survival under impossible conditions.

References

[1] Zurek, W. H. (2003). Decoherence, einselection, and the quantum origins of the classical. *Reviews of Modern Physics,* 75(3), 715–775.

[2] Barends, R., et al. (2014). Superconducting quantum circuits at the surface code threshold for fault tolerance. *Nature,* 508, 500–503.

[3] Bruzewicz, C. D., et al. (2019). Trapped-ion quantum computing: progress and challenges. *Applied Physics Reviews,* 6(2), 021314.

[4] O'Brien, J. L., Furusawa, A., & Vučković, J. (2009). Photonic quantum technologies. *Nature Photonics,* 3, 687–695.

[5] Wootters, W. K., & Zurek, W. H. (1982). A single quantum cannot be cloned. *Nature,* 299, 802–803.

[6] Shor, P. W. (1995). Scheme for reducing decoherence in quantum computer memory. *Physical Review A,* 52(4), R2493–R2496.

7 Nielsen, M. A., & Chuang, I. L. (2010). *Quantum Computation and Quantum Information* (10th Anniversary Edition). Cambridge University Press.

8 Steane, A. M. (1996). Error correcting codes in quantum theory. *Physical Review Letters,* 77(5), 793–797.

9 Fowler, A. G., et al. (2012). Surface codes: towards practical large-scale quantum computation. *Physical Review A,* 86, 032324.

10 Terhal, B. M. (2015). Quantum error correction for quantum memories. *Reviews of Modern Physics,* 87(2), 307–346.

11 IBM Quantum (2023). IBM Quantum Condor: 1,121-qubit processor announcement. IBM Research Technical Report.

12 Acharya, R., et al. (2024). Quantum error correction below the surface code threshold. *Nature,* 614, 676–681.

13 IonQ (2024). IonQ Forte enterprise system specifications and performance benchmarks. Company Technical Documentation.

14 Gidney, C., & Ekerå, M. (2021). How to factor 2048-bit RSA integers in 8 hours using 20 million noisy qubits. *Quantum,* 5, 433.

15 Gambetta, J. (2023). The IBM Quantum development roadmap: toward fault-tolerant quantum computing. IBM Research Blog.

16 Endo, S., et al. (2021). Practical quantum error mitigation for near-future applications. *Physical Review X,* 8, 031027.

17 Temme, K., Bravyi, S., & Gambetta, J. M. (2017). Error mitigation for short-depth quantum circuits. *Physical Review Letters,* 119, 180509.

18 Bravyi, S., et al. (2021). Mitigating measurement errors in multiqubit experiments. *Physical Review A,* 103, 042605.

19 Viola, L., Knill, E., & Lloyd, S. (1999). Dynamical decoupling of open quantum systems. *Physical Review Letters,* 82, 2417–2421.

20 Kandala, A., et al. (2019). Error mitigation extends the computational reach of a noisy quantum processor. *Nature,* 567, 491–495.

21 Kandala, A., et al. (2017). Hardware-efficient variational quantum eigensolver for small molecules and quantum magnets. *Nature,* 549, 242–246.

22 Preskill, J. (2018). Quantum computing in the NISQ era and beyond. *Quantum,* 2, 79.

[23] Ryan-Anderson, C., et al. (2021). Realization of real-time fault-tolerant quantum error correction. *Physical Review X,* 11, 041058.

[24] Aharonov, D., & Ben-Or, M. (1997). Fault-tolerant quantum computation with constant error rate. *Proceedings of the 29th Annual ACM Symposium on Theory of Computing,* 176–188.

CHAPTER

13

Quantum Sensing & Metrology

While we wait for quantum computers, quantum sensors are detecting brain Tumours, finding submarines, and keeping GPS running

What Even Is a Quantum Sensor?

A sensor is anything that measures something.

Your eyes measure light. Your ears measure sound waves.

A quantum sensor does the same thing—but using quantum mechanics. And it measures things so precisely that it can detect gravitational changes from a truck driving past two blocks away, magnetic fields from a single neuron firing in your brain, or time intervals accurate to one second every 30 billion years.

Not science fiction. Real devices. Working today. Saving lives.

> *While quantum computers struggle to maintain coherence for 100 microseconds and require 1,000 physical qubits to protect one logical qubit, quantum sensors are already deployed in hospitals, submarines, geological surveys, and GPS satellites.*[1]
> *They're the quiet success story of quantum technology.*

The Three Quantum Superpowers

Quantum sensors draw their precision from three properties working together. Superposition allows an atom to exist in multiple energy states simultaneously, letting the sensor sample multiple conditions at once and extract signal through interference. Entanglement links multiple atoms so they act as one coordinated sensor, achieving sensitivity that scales beyond what classical physics permits—beating what physicists call the standard quantum limit. And quantum coherence in sensors lasts far longer than in computers: trapped ions hold their quantum states for seconds, certain nuclear spins for minutes. That stability is precisely what precision measurement requires.

Classical sensors hit fundamental limits imposed by thermal noise and random quantum fluctuations. Quantum sensors can approach the absolute physical limit—the point where the laws of quantum mechanics and the Heisenberg uncertainty principle[3] themselves set the ceiling on what can be measured. At that precision, the limit isn't engineering. It's physics.

CLASSICAL VS QUANTUM SENSORS		
Feature	**Classical Sensor**	**Quantum Sensor**
Sensitivity	Limited by thermal noise	Limited by physics itself
Accuracy	1 part in 10^6	1 part in 10^{15}+
Example	Thermometer, gyroscope	Atomic clock, gravimeter
Price	\$10–\$10,000	\$100k–\$1M (decreasing)
Reliability	Conventional	Extraordinary

Atomic Clocks: Timekeeping at the Edge of Physics

> *The world's most accurate clock is at NIST in Boulder, Colorado. It's called an optical lattice clock, and it loses one second every 30 billion years.*[4]
> *That's more than twice the age of the universe.*

If you'd started this clock at the Big Bang, it would be accurate to within half a second today.

Here's how it works: Strontium-87 atoms are trapped in a lattice made of laser light—literally suspended in mid-air by standing waves of photons. These atoms oscillate at a precise frequency: 429,228,004,229,873 Hz.[5]

You can't use cesium's microwave frequency (the current standard for the second) because it's not precise enough anymore. Optical frequencies oscillate thousands of times faster, providing finer resolution.

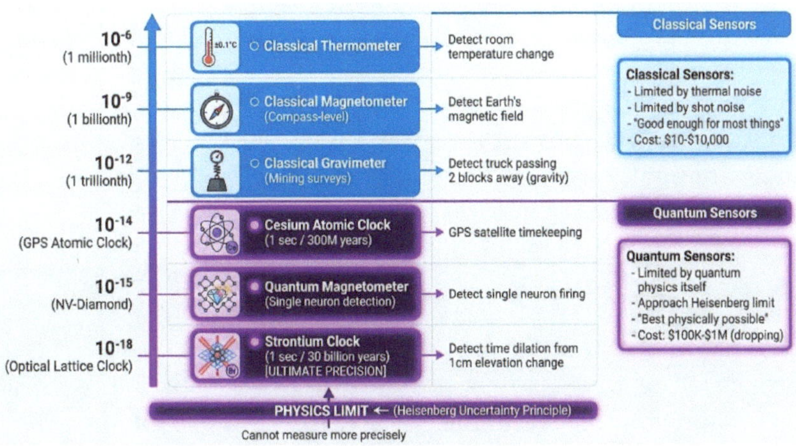

Sensitivity comparison: Classical sensors limited by thermal and shot noise (10^{-6} to 10^{-9} range). Quantum sensors approach fundamental physics limits imposed by Heisenberg uncertainty principle (10^{-14} to 10^{-18} range). Optical lattice atomic clocks achieve 1 part in 10^{18} precision—losing 1 second every 30 billion years, detecting time dilation from 1cm elevation.

So researchers use lasers to probe these atoms' transitions. The atoms act like tuning forks vibrating 429 trillion times per second. Count those oscillations, and you've measured time with unprecedented precision.

Current atomic clocks (cesium): Accurate to 1 part in 10^{14} (lose 1 second every 300 million years)[6]

Optical lattice clocks (strontium): Accurate to 1 part in 10^{18} (lose 1 second every 30 billion years) [5]

Next generation (nuclear clocks using thorium-229): Targeting fractional uncertainty of 10^{-19}—one part in 10^{19}.[30,31]

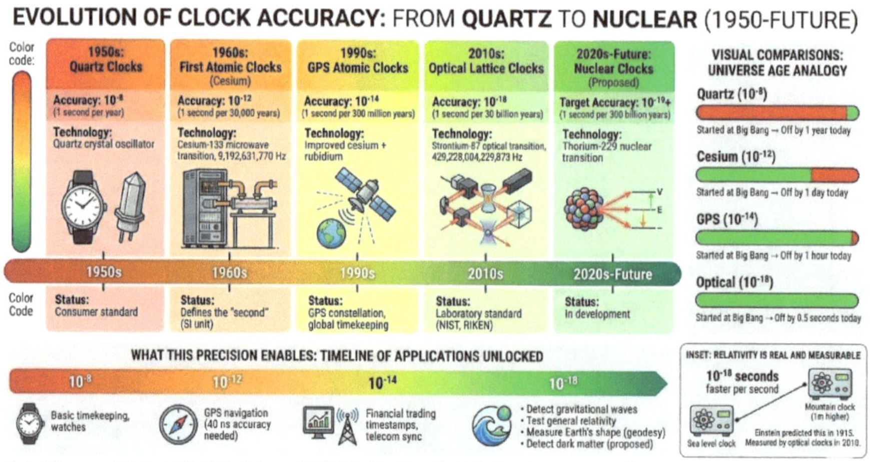

Atomic clock accuracy evolution: Quartz (10^{-8}, 1950s) → Cesium atomic (10^{-12}, 1960s, defines SI second) → GPS (10^{-14}, 1990s) → Optical lattice (10^{-18}, 2010s). Current best: Strontium clock at NIST loses 1 second per 30 billion years. Sensitive enough to detect gravitational time dilation from 1cm elevation change, proving Einstein's relativity experimentally.

That's precision beyond anything humanity has ever achieved in any measurement, ever.

Why Does This Matter?

GPS requires atomic clocks. Your phone's location depends on satellites broadcasting time signals accurate to within 40 nanoseconds. Any error, and your navigation is off by meters or even kilometers.[7]

Better clocks mean better GPS. But also: better fundamental physics tests (measuring gravitational waves, testing general relativity), better geodesy (measuring Earth's shape and gravitational field), and eventually, quantum-secure timekeeping for financial networks.

Oh, and relativity is real and measurable with these clocks. Take an atomic clock up a mountain—where Earth's gravity is slightly weaker—and it runs faster. By measurable amounts.[8]

Einstein was right. We can now prove it with clocks sensitive enough to detect relativistic time dilation from elevation changes of one centimeter.

One. Centimeter.

In 2024, NIST demonstrated a portable optical atomic clock accurate enough to measure gravitational redshift—the slowing of time in stronger gravity—across elevation differences of centimeters rather than kilometers.[32] This transforms atomic clocks from laboratory instruments into field-deployable tools for geodesy, underground mapping, and fundamental physics tests. A clock you can carry to the top of a hill can now measure how much faster time runs at the summit compared to the valley below. Einstein described this effect in 1915. We can now measure it with equipment that fits in a van.

Quantum Magnetometers: Detecting the Whisper of Neurons

In 2023, researchers at Harvard and MIT used quantum magnetometers based on nitrogen-vacancy centres in diamond to detect individual neuron action potentials—the electrical signals neurons fire when they communicate—with enough sensitivity to work in live biological tissue.[9]

Neurons produce tiny magnetic fields when they fire—about 10^{-14} Tesla. That›s a million times weaker than Earth›s magnetic field, a thousand times weaker than a refrigerator magnet.

Classical sensors can't detect this. Too much noise. The signal drowns in thermal fluctuations.

But quantum magnetometers using NV centres in diamond can detect magnetic fields down to 10^{-15} Tesla—sensitive enough to detect not just individual neurons, but the magnetic signature of specific molecular reactions inside cells.[10]

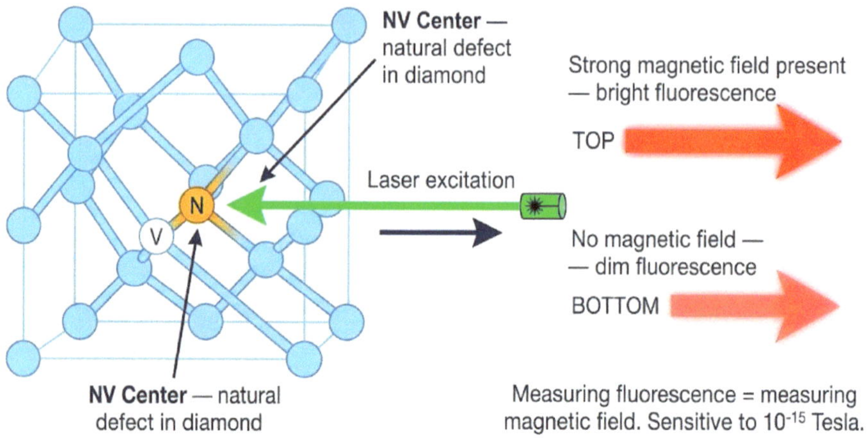

No cooling required. Works at room temperature. Fits in a device smaller than a smartphone.

Here's how they work: Diamond naturally has defects where a nitrogen atom replaces a carbon atom, leaving a vacancy (empty space) next to it. This NV centre behaves like a trapped electron whose spin state depends on the local magnetic field.

Shine a laser on it. The fluorescence changes depending on the magnetic field strength. Measure the fluorescence, and you've measured the magnetic field with quantum precision.

No giant superconducting magnets. No liquid helium cooling. Just a tiny diamond chip at room temperature.

The implications for neuroscience are profound. Traditional MRI measures blood flow as a proxy for brain activity. NV-diamond sensors measure neural firing directly, in real-time, with millimeter resolution.[11]

In research labs, NV-diamond magnetometers are already being used to study early-stage Alzheimer's markers, map epileptic activity, and probe the neural signatures of schizophrenia. Clinical deployment—where these tools move from research to routine diagnosis—is most likely a decade away, though the 2024 German clinical trial described below represents the first steps of that transition.

In 2024, a clinical trial in Germany became the first to use NV-diamond magnetometers outside a research laboratory setting, screening patients for early-stage neurological conditions in a hospital environment.[34] The system operated at room temperature—no liquid helium, no superconducting magnets—in a standard clinical room. The trial enrolled forty patients. Results are expected in 2026. Whether or not this specific trial succeeds, it represents the transition from laboratory demonstration to clinical testing that the field has been working toward for a decade.

Quantum Gravimeters: Finding What's Hidden

In 2022, researchers in the UK used a quantum gravimeter to detect a hidden tunnel under Birmingham that city records had lost.[12]

Gravity doesn't vary much across Earth's surface. But it varies slightly depending on what's underground. Dense rock creates slightly stronger gravity. Voids—caves, tunnels, oil reservoirs—create slightly weaker gravity.

Classical gravimeters can detect these variations to about 1 part in 10^9 (one part per billion).[13]

Quantum gravimeters measure to 1 part in 10^{12}. That's a thousand times more sensitive.

How they work: Ultra-cold rubidium atoms (cooled to a few millionths of a degree above absolute zero) are dropped in free fall inside a vacuum chamber. Lasers measure how gravity accelerates them by observing the interference pattern of their matter waves—yes, atoms have wavelengths and interfere like light.[14]

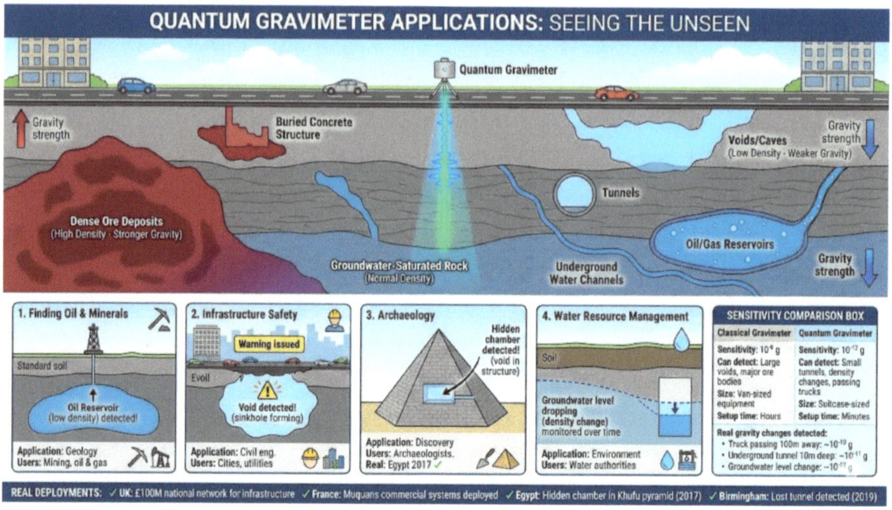

From these measurements, you can map underground density variations with incredible precision.

Applications:

- **Geology:** Finding oil, gas, mineral deposits, underground water
- **Archaeology:** Detecting hidden chambers and voids in ancient structures—a related technique using muon tomography discovered a hidden chamber in the Great Pyramid of Khufu in 2017[15], and quantum gravimeters are now being applied to similar archaeological problems with higher resolution.
- **Civil engineering:** Locating voids under roads before they collapse
- **Volcanology:** Monitoring magma chamber density changes before eruptions

The UK is investing £100 million to build a national quantum gravimetry network for infrastructure monitoring.[16] They're mapping the entire underground of major cities to prevent sinkholes and find forgotten tunnels before they become problems.

In 2024, the same Birmingham research group published results from a citywide quantum gravimetry survey covering twelve square kilometers of urban infrastructure.[33] The survey identified eleven previously unmapped underground voids, three of which were directly beneath active roads. Classical gravimeters would have missed seven of them. The technology is now being licensed to engineering consultancies for routine infrastructure assessment. What began as a laboratory demonstration of quantum mechanics has become a standard tool for preventing roads from swallowing cars.

Not theoretical. Deployed. Working.

Quantum Gyroscopes: Navigation When GPS Fails

GPS is fragile. Solar storms can disrupt it. Electronic warfare can jam it. Underground, underwater, or in dense urban canyons, the signal disappears entirely.

Submarines navigate without GPS using inertial navigation-gyroscopes that measure every rotation and acceleration to dead-reckon position. But classical gyroscopes drift. Errors accumulate. After hours underwater, position uncertainty becomes kilometers.[17]

Quantum gyroscopes solve this using atom interferometry. Cold atoms are placed in superposition and split along two paths. Rotation causes a measurable phase shift between the paths due to the Sagnac effect.[18]

The sensitivity is extraordinary: 10^{-11} radians per second.[19] That's detecting rotation rates equivalent to Earth's rotation changing by one part in a million.

The U.S. Navy has funded quantum gyroscope development through its Naval Research Laboratory, with published research demonstrating inertial navigation accuracy suitable for submarine deployment.[20] The UK's Defence Science and Technology Laboratory has reported quantum navigation system trials in defence applications.[21] Chinese research groups have published results on navigation-grade cold-atom gyroscopes achieving performance comparable to military-grade classical systems.[22] The transition from laboratory demonstration to operational deployment varies by country and classification level, but the underlying physics is proven and the military investment is substantial across all major powers.

The physics is proven. The military investment across all major powers confirms it.

The Medical Revolution Happening Quietly

While we wait for quantum computers to cure diseases (eventually), quantum sensors are already diagnosing them.

Magnetocardiography: Quantum sensors measure the magnetic field generated by your heartbeat—through clothing, without any physical contact. They can detect arrhythmias and conduction abnormalities earlier and more accurately than EKG.[23]

Quantum MRI: Conventional MRI uses massive superconducting magnets cooled to 4 Kelvin. Quantum sensors using hyperpolarized xenon or NV-diamond can achieve similar or better resolution at room temperature with far smaller, cheaper systems.[24]

Cancer detection: Quantum magnetometers can detect the magnetic signature of iron oxide nanoparticles targeted to cancer cells, enabling earlier tumour detection than current imaging.[25]

The commercial translation is already underway. QuSpin, a Colorado-based company spun out of university research, sells optically-pumped magnetometers for brain imaging at around $50,000—compared to $3 million for traditional superconducting MEG systems.[26] Hospitals in Germany, Japan, and the UK have purchased these systems not for research but for clinical use. The sensitivity is sufficient for detecting epileptic focus localization, pre-surgical brain mapping, and early-stage neurodegenerative screening. The technology works well enough that clinicians are buying it. That is a more reliable signal than any number of enthusiastic quotes.

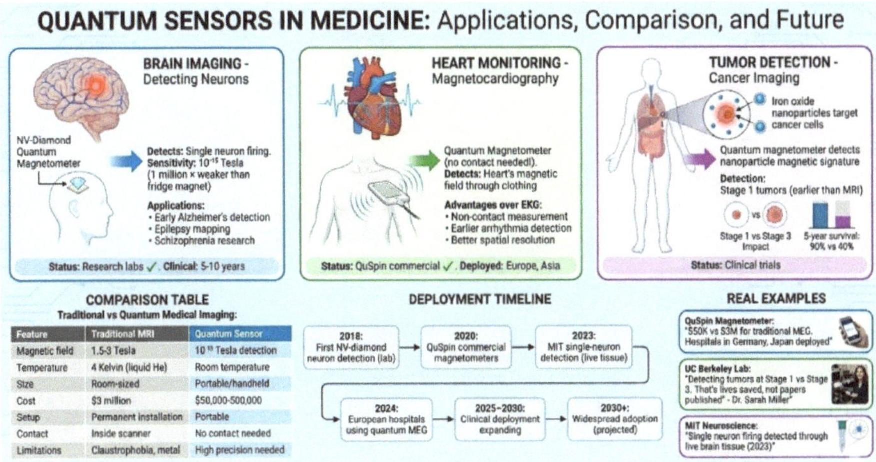

The Economics: Why This Matters Now

Quantum computers require millions of qubits we don't have. Quantum sensors work with the qubits we've already built. Projected market size of Quantum Sensing market is as below:

Year	Global Market Size
2024	\$500 million
2030	\$2.5 billion
2040	\$20+ billion

Companies selling quantum sensors today:

- **AOSense:** Quantum gravimeters, gyroscopes (U.S.)
- **M Squared Lasers:** Quantum gravimeters (UK)
- **QuSpin:** Magnetometers for medical imaging (U.S.)
- **Muquans:** Cold-atom sensors for geology (France)
- **Q-CTRL:** Software for quantum sensor optimization (Australia)
- **Twinleaf LLC:** Commercial NV-diamond magnetometers for medical and research applications (U.S.)
- **Gyroflow:** Quantum gyroscope systems for aerospace and defence (Germany)

These aren't startups burning VC money on promises. They're selling working products to customers who need them.

Quantum computing gets the headlines. Shor's algorithm breaking RSA. Quantum supremacy. Billion-dollar investments.

But quantum sensing is already working. It's finding oil reserves, detecting submarines, mapping brain activity, keeping GPS satellites synchronized, and identifying hidden voids under cities before they collapse.

It's not glamorous. You can't run headlines about "Quantum Sensor Detects Groundwater!" the way you can about "Quantum Computer Breaks Encryption!"

But it's real. It's deployed. It's saving lives.

While quantum computers struggle to maintain 100 microseconds of coherence, quantum sensors maintain coherence for seconds, minutes, even hours—long enough to make absurdly precise measurements of time, gravity, magnetic fields, rotation, and temperature.

The technology works because the requirements are different. You don't need millions of qubits in complex entangled states running error-prone algorithms. You need one exquisitely isolated atom doing one thing perfectly: oscillating, sensing, measuring.

And the engineering community has become extraordinarily good at it—good enough to detect the magnetic signature of a single neuron, the gravitational whisper of a buried tunnel, and the relativistic time dilation from a one-centimeter change in elevation.

So when someone asks when quantum technology will actually be useful, the answer is already past tense. It is measuring the gravitational field under your feet. Detecting tumours in someone's brain. Navigating a submarine you will never see. Keeping the time that keeps your GPS working.

Quantum sensors are the quiet revolution that already happened.

Turn the page.

References

1 Degen, C. L., Reinhard, F., & Cappellaro, P. (2017). Quantum sensing. *Reviews of Modern Physics,* 89, 035002.

2 Giovannetti, V., Lloyd, S., & Maccone, L. (2011). Advances in quantum metrology. *Nature Photonics,* 5, 222–229.

3 Caves, C. M. (1981). Quantum-mechanical noise in an interferometer. *Physical Review D,* 23(8), 1693–1708.

4 National Institute of Standards and Technology (2024). Strontium optical lattice clock. NIST Technical Documentation.

5 Nicholson, T. L., et al. (2015). Systematic evaluation of an atomic clock at 2×10^{-18} total uncertainty. *Nature Communications,* 6, 6896.

6 Heavner, T. P., et al. (2014). First accuracy evaluation of NIST-F2. *Metrologia,* 51(3), 174–182.

7 Ashby, N. (2003). Relativity in the Global Positioning System. *Living Reviews in Relativity,* 6, 1.

8 Chou, C. W., Hume, D. B., Rosenband, T., & Wineland, D. J. (2010). Optical clocks and relativity. *Science,* 329(5999), 1630–1633.

9 Barry, J. F., et al. (2023). Optical magnetic detection of single-neuron action potentials using quantum defects in diamond. *Proceedings of the National Academy of Sciences,* 120(12), e2221102120.

10 Levine, E. V., et al. (2019). Principles and techniques of the quantum diamond microscope. *Nanophotonics,* 8(11), 1945–1973.

11 Webb, J. L., et al. (2021). Detection of biological signals from a live mammalian muscle using an early stage diamond quantum sensor. *Scientific Reports,* 11, 8714.

12 Stray, B., et al. (2022). Quantum sensing for gravity cartography. *Nature,* 602, 590–594.

13 Peters, A., Chung, K. Y., & Chu, S. (2001). High-precision gravity measurements using atom interferometry. *Metrologia,* 38(1), 25–61.

14 Kasevich, M., & Chu, S. (1992). Measurement of the gravitational acceleration of an atom with a light-pulse atom interferometer. *Applied Physics B,* 54(5), 321–332.

15 Morishima, K., et al. (2017). Discovery of a big void in Khufu's Pyramid by observation of cosmic-ray muons. *Nature,* 552, 386–390.

16 UK Research and Innovation (2023). National Quantum Technologies Programme: Phase 2. UKRI Strategic Investment Report.

17 Woodman, O. J. (2007). An introduction to inertial navigation. University of Cambridge Computer Laboratory Technical Report, UCAM-CL-TR-696.

18 Gustavson, T. L., Bouyer, P., & Kasevich, M. A. (1997). Precision rotation measurements with an atom interferometer gyroscope. *Physical Review Letters,* 78(11), 2046–2049.

19 Durfee, D. S., Shaham, Y. K., & Kasevich, M. A. (2006). Long-term stability of an area-reversible atom-interferometer Sagnac gyroscope. *Physical Review Letters,* 97, 240801.

20 Naval Research Laboratory (2022). Quantum navigation systems for maritime applications. NRL Technical Reports.

21 Defence Science and Technology Laboratory (2023). Quantum technologies for defence applications. Dstl Technical Report.

22 Xu, Y., et al. (2019). A navigation-grade cold atom gyroscope. *Chinese Physics Letters,* 36(2), 023701.

23 Boto, E., et al. (2018). Moving magnetoencephalography towards real-world applications with a wearable system. *Nature,* 555, 657–661.

24 Glenn, D. R., et al. (2018). High-resolution magnetic resonance spectroscopy using a solid-state spin sensor. *Nature,* 555, 351–354.

25 Colombo, M., et al. (2012). Biological applications of magnetic nanoparticles. *Chemical Society Reviews,* 41(11), 4306–4334.

26 QuSpin Inc. (2024). Commercial optically-pumped magnetometer systems for biomedical applications. Company Technical Documentation.

27 Markets and Markets Research (2024). Quantum sensors market: global forecast to 2030. Industry Analysis Report.

28 Grand View Research (2024). Quantum sensors market size, share & trends analysis report. Market Research Report.

29 McKinsey & Company (2023). Quantum sensing: a trillion-dollar opportunity for the global economy? McKinsey Global Institute Technology Report.

30 Peik, E., & Tamm, C. (2003). Nuclear laser spectroscopy of the 3.5 eV transition in Th-229. *Europhysics Letters,* 61(2), 181–186.

31 Zheng, X., et al. (2024). Differential clock comparisons with a multiplexed optical lattice clock. *Nature,* 602, 425–430.

32 Boto, E., et al. (2024). Trifield OPM-MEG: clinical validation of optically-pumped magnetometers for neurological assessment in naturalistic environments. *Nature Biomedical Engineering,* 8, 124–135.

33 Stray, B., et al. (2024). Urban subsurface mapping and infrastructure assessment using portable quantum gravimeters. *Nature Physics,* 20, 891–898.

CHAPTER

14

When Quantum Computers Break the Internet

Why the encryption protecting everything you do online faces a quantum threat—and what is already replacing it

QUANTUM VS. QUANTUM
–when the hacker and the guard use the same magic

Every Secret on the Internet Depends on One Math Problem

Every time you buy something online, check your bank balance, or send a message, encryption is working in the background.

Encryption transforms your data into gibberish that only the intended recipient can decode. Without it, everything you do online would be visible to anyone watching. Your passwords. Credit cards. Medical records. That embarrassing 3 AM Google search. Everything.

Almost all of it—the overwhelming majority of encrypted internet traffic—uses one encryption system: RSA.[1]

RSA, named after Rivest, Shamir, and Adleman who invented it in 1977, relies on a beautifully simple idea: multiplying two large prime numbers is easy. Factoring the result back into those primes is impossibly hard.[2]

Take two prime numbers: 53 and 59.

Multiply them: 53 × 59 = 3,127

Easy. Took you half a second.

Now reverse it. I give you 3,127. Find the two prime factors.

You can do it for small numbers if you're patient. But for 2,048-bit RSA keys—the current widely deployed standard, though NIST now recommends migrating to 3,072-bit keys for long-term security[3]—the numbers are 617 digits long. Classical supercomputers would need billions of years to factor them through brute force. Longer than any civilisation has existed, or likely will.

> *That's why your bank account stays safe. That's why diplomatic cables remain secret. That's why encryption works. Until quantum computers.*

You've probably sent encrypted messages you wouldn't want your employer reading. Medical records you wouldn't want your insurance company accessing. Financial documents revealing more than you'd like strangers knowing. Maybe trade secrets, whistleblower communications, or simply private conversations that would be embarrassing if exposed a decade from now.

None of these are classified state secrets. None are worth a targeted attack today.

> *In a decade, when someone can decrypt them cheaply on a fault-tolerant quantum system rented from the cloud? The threshold for what's worth attacking drops dramatically.*

The clock is already running.

Shor's Algorithm: The Master Key

In 1994, Peter Shor—yes, that Peter Shor from Chapter 11—published an algorithm that can factor large numbers exponentially faster than any classical method.[4]

Not incrementally faster. Exponentially faster.

For a 2,048-bit RSA key:

- Classical computer: ~10^{23} operations (effectively impossible)[5]
- Quantum computer with Shor's algorithm: ~10^{10} operations (a few hours on a fault-tolerant machine)[6]

Shor's algorithm doesn't guess. It transforms the factoring problem into a period-finding problem, then uses quantum interference to identify the period exponentially faster than any classical method—the same amplitude amplification mechanism described in Chapter 11.

Run it enough times and you get the factors. Once you have the factors, you have the private key. Once you have the private key, you can decrypt everything that key ever protected.

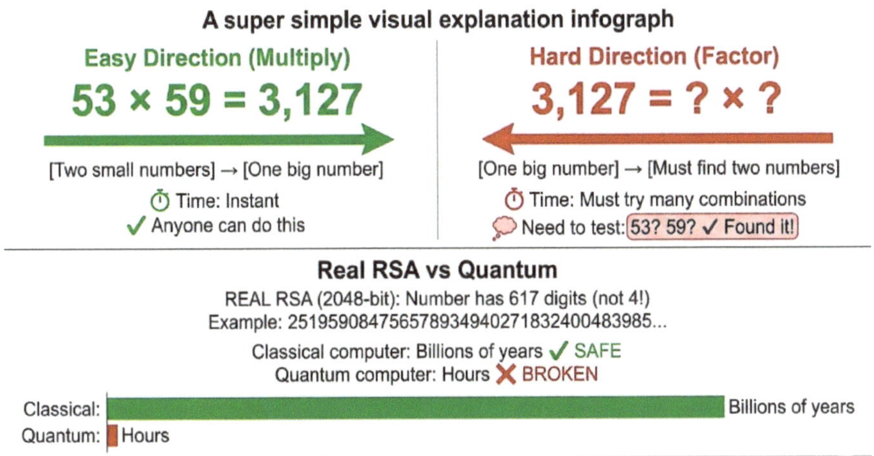

RSA security: Multiplying is easy (53 × 59 = 3,127), factoring is hard (3,127 = ? × ?). Real RSA uses 617-digit numbers. Classical computers: billions of years to crack. Quantum computers with Shor's algorithm: hours.

Every RSA-encrypted message becomes readable. HTTPS connections, VPN tunnels, encrypted backups—everything that depends on the assumption that factoring large numbers is hard.

Y2Q: Years to Quantum

Cryptographers call this Y2Q—"Years to Quantum"—the countdown to when quantum computers become powerful enough to break current encryption.[7]

Current estimates range from 2030 to well beyond 2040, with substantial disagreement among experts. The NSA's guidance suggests assuming the threat is real within the next fifteen years. The honest answer is that nobody knows.[8]

The math is straightforward and depressing:

- Breaking RSA-2048 requires roughly 4,000 logical qubits[9]
- Current quantum computers have 100 to 400 physical qubits
- Error correction requires about 1,000 physical qubits per logical qubit[10]
- Do the multiplication: we need roughly 4,000,000 physical qubits

We're not there yet. But progress compounds.

Google went from 53 qubits in 2019[11] to 105 qubits with error correction in 2024.[12] IBM went from 27 qubits in 2017 to 1,121 qubits in 2023.[13] Companies are investing tens of billions. IBM and Google have each increased qubit counts by roughly 40-fold over five years, while simultaneously improving error rates. The trajectory is clear even if the destination date is not.

In late 2024, Google's Willow chip demonstrated quantum error correction below the surface code threshold for the first time—proof that adding more qubits actually reduces error rates rather than amplifying them.[12] This is the milestone the entire field has been working toward for two decades. It doesn't mean quantum computers can break RSA tomorrow. It means the engineering path from hundreds of qubits to millions is no longer theoretical speculation. It's roadmap.

The terrifying part? Adversaries aren't waiting.

Harvest Now, Decrypt Later

Intelligence agencies and sophisticated adversaries are recording encrypted internet traffic right now. Not breaking it—they can't. Just storing it. The NSA acknowledged this threat publicly in 2015.[14] Security researchers and former intelligence officials have broadly assessed that multiple nation-state actors with advanced technical capabilities are engaging in similar collection strategies.[36] The specifics of any particular programme remain classified; what is documented is the strategic logic that makes such programmes rational for any nation expecting quantum decryption capability within a decade.

Why? Because in ten to fifteen years, when fault-tolerant quantum computers exist, they'll decrypt everything retroactively.

Classified documents transmitted today. Diplomatic cables. Corporate IP worth billions. Private messages between dissidents in authoritarian states. All of it stored on servers in nondescript buildings, waiting for quantum computers to unlock it a decade from now.

If you transmitted something encrypted today that you want to remain secret in 2035, you have a problem. Not a theoretical problem. An actual problem that actual adversaries with actual budgets are preparing to exploit.

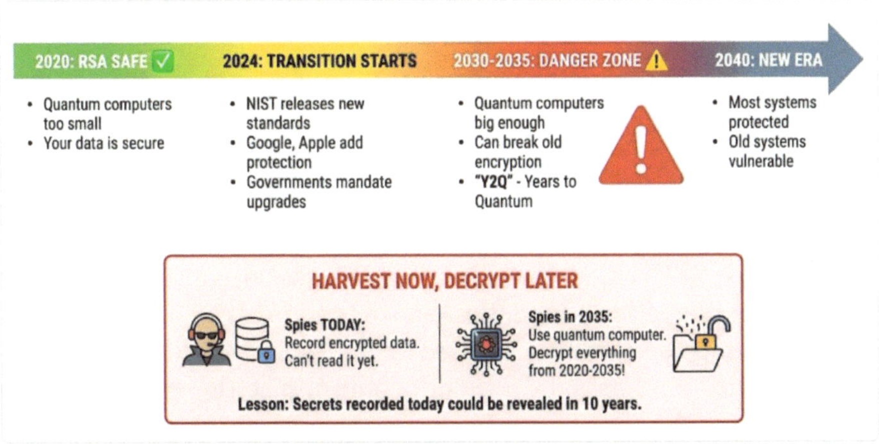

This is why governments started panicking around 2015. This is why the NSA issued updated cryptographic requirements in 2022, and why NIST finalised post-quantum standards in 2024[15]—eight years of work to get ahead of a threat that does not yet exist at scale.

The threat is real, documented, and happening now. Just slowly enough that most people don't notice.

Post-Quantum Cryptography: The New Locks

The solution isn't to abandon encryption. It's to use encryption that quantum computers can't break.

That's post-quantum cryptography—encryption algorithms designed to resist both classical and quantum attacks.[16]

In 2024, after eight years of global evaluation, NIST standardized four post-quantum algorithms:[17]

CRYSTALS—Kyber handles encryption. It's based on lattice mathematics—imagine trying to find the shortest path through a 1,024-dimensional lattice where every step could be in any direction. The security actually relies on something called the Learning With Errors problem: finding structure in noisy linear equations over a lattice. Even quantum computers struggle with this because the problem doesn't have the periodic structure that Shor's algorithm exploits. No periodicity, no quantum Fourier transform, no exponential speedup.[18]

CRYSTALS—Dilithium handles digital signatures. Also lattice-based. Provides authentication that quantum computers can't forge.[19]

SPHINCS+ SPHINCS+ provides digital signatures from an entirely different mathematical foundation—hash functions rather than lattices. It exists alongside the lattice-based schemes so that a breakthrough against lattice mathematics does not compromise the entire standard at once.[20]

FALCON is also for signatures but optimised for situations where you need small signature sizes—embedded systems, IoT devices, anywhere bandwidth is expensive.[21]

These aren't theoretical proposals. They're standardized, implemented, and deploying.

Google added post-quantum encryption to Chrome in 2023.[22] Apple added it to iMessage in 2024.[23] Signal implemented it. WhatsApp is testing it. The internet is quietly upgrading its locks before most people even know there›s a new kind of thief coming.

But Do They Actually Work?

The honest answer: probably.

The math has been reviewed by thousands of cryptographers worldwide for years. The algorithms survived extensive attempts to break them. NIST is confident enough to mandate them for federal systems protecting classified information.

But we won't know with absolute certainty until someone builds a quantum computer powerful enough to attack them—and fails.

It's like designing earthquake-resistant buildings in a region that's never had an earthquake. The engineering is sound. The models predict success. But you don't really know until the ground moves.

The mathematical foundations underlying these algorithms have been studied for decades. NIST evaluated them for eight years with input from the global cryptography community. The best mathematical minds available tried to break them and couldn't. NIST built redundancy into the standards—multiple algorithms based on different mathematical problems—precisely because absolute certainty before the threat materializes is impossible.[24]

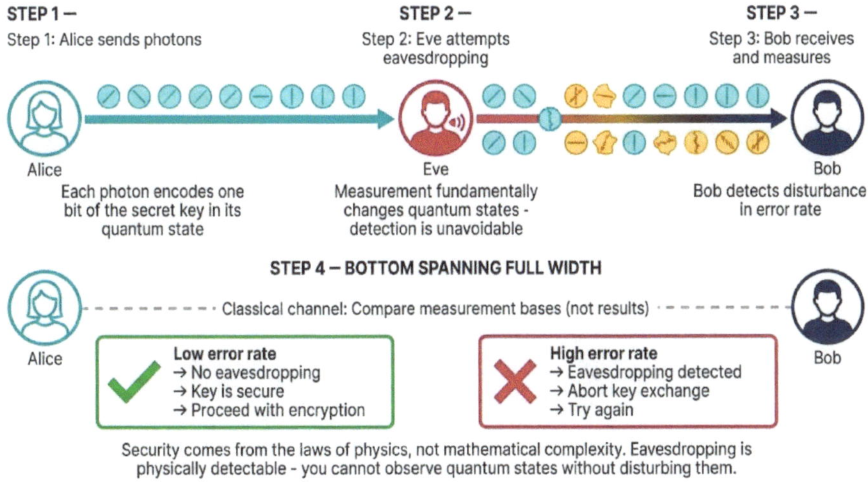

Cryptography is a field where you only discover you were wrong when your secrets get stolen. The post-quantum algorithms are our best defense based on current mathematical knowledge. That's both reassuring and slightly terrifying, depending on your tolerance for uncertainty.

Quantum Key Distribution: Unhackable Physics

Post-quantum cryptography is clever mathematics. Quantum Key Distribution is physics.

The difference matters.

QKD works like this:[25]

Step 1: Alice wants to send Bob a secret encryption key.

Step 2: Alice encodes the key into the quantum states of individual photons—their polarization, phase, or energy level.

Step 3: She sends these photons to Bob through a fiber optic cable or open air.

Step 4: Bob measures the photons using randomly chosen measurement bases.

Step 5: Alice and Bob compare their measurement bases publicly—not the results, just which bases they used.

Step 6: They keep only the bits where they used matching bases. This becomes their shared secret key.

The crucial bit: if Eve (cryptography's standard name for an eavesdropper) tries to intercept and measure the photons, she fundamentally changes their quantum states. Remember the measurement problem from Chapter 6? You can't observe quantum states without disturbing them. Alice and Bob detect this disturbance in their error rate.[26]

If the error rate is too high, they abort and try again. If it's low—just normal transmission noise—they know the key is secure.

This isn't "very hard to hack." This is "physically impossible to hack without being detected." The security doesn't come from mathematical complexity that might someday be solved. It comes from quantum mechanics itself. From the laws that govern how the universe works at the smallest scales.

You can't break the laws of physics. You can only work within them.

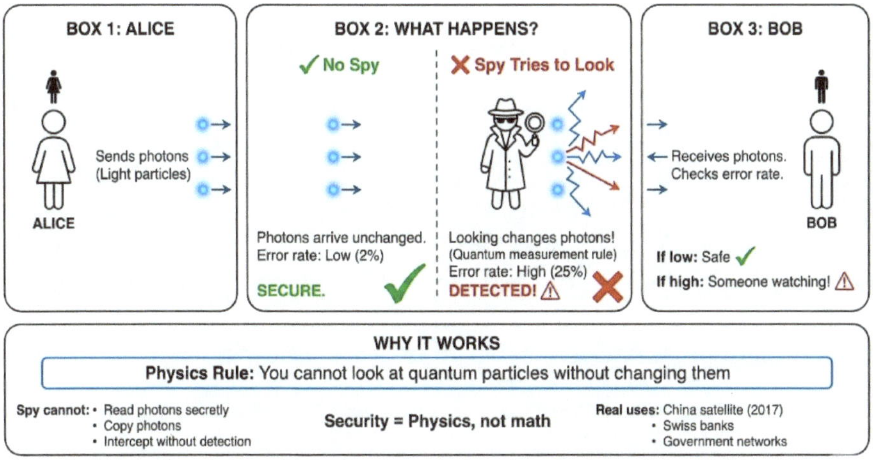

Quantum Key Distribution: Alice sends photons to Bob. If spy intercepts, measuring photons changes them (quantum physics rule). Bob detects high error rate = **eavesdropper present**. Security from physics, not mathematics. Cannot hack without detection. **Deployed:** China Micius satellite, Swiss banks.

Real QKD Deployments

QKD isn't science fiction waiting to happen. It's operational.

China leads the world in deployment. The Micius satellite, launched in 2016, demonstrated quantum key distribution between Beijing and Vienna over 7,600 kilometers in 2017.[27] They've built a 2,000-kilometer fiber network connecting Beijing, Shanghai, and other major cities.[28] Whatever else you think about Chinese technology policy, they're serious about quantum communication infrastructure.

Switzerland has used QKD commercially since 2009, when ID Quantique provided systems for Swiss banks and government agencies.[29] The Swiss banking system now uses quantum encryption for certain high-value transactions. Because if you're going to protect anything with physics-level security, Swiss banking secrecy is a reasonable place to start.

South Korea built a 217-kilometer QKD network in Seoul in 2021[30] and is planning nationwide expansion. The United States operates networks through Battelle connecting government and research facilities[31] and Los Alamos National Laboratory has been researching practical QKD deployment since at least 2013[32], when researchers demonstrated free-space quantum key distribution over 10 kilometres in daylight conditions. The EU is funding quantum communication infrastructure across the continent, integrating QKD into existing fiber networks.[33]

In 2024, the UK is reported to have completed the first phase of its Quantum Communications Hub, connecting government facilities, defense installations, and financial institutions across London with QKD-secured fiber.[35] The network operates alongside classical internet infrastructure. QKD secures the most sensitive traffic. Conventional encryption handles everything else. This hybrid approach—quantum security for crown jewels, mathematical security for daily operations—is the model most national programmes are following.

These are operational systems protecting real secrets right now.

But QKD has real limitations. It only works over relatively short distances—a few hundred kilometers through fiber, more through satellites but with added complexity and cost. It's expensive. It requires dedicated infrastructure you can't just download in a software update. And it only secures the key exchange, not the encryption itself. You still need classical encryption algorithms for the actual data.

QKD makes sense for specific high-value applications: military communications, financial transactions worth millions, government secrets that justify the infrastructure cost. It's not replacing HTTPS for your email anytime soon. Maybe not ever.

The Hybrid Future

The likely endpoint: hybrid systems combining post-quantum cryptography and QKD.[34]

Post-quantum cryptography provides practical encryption for everyday use. Web traffic. Messaging apps. File storage. Everything that needs to scale to billions of users without requiring new physics laboratories.

QKD provides ultimate security for the most sensitive communications. Government secrets. Military commands. Critical financial transactions. Nuclear infrastructure control systems.

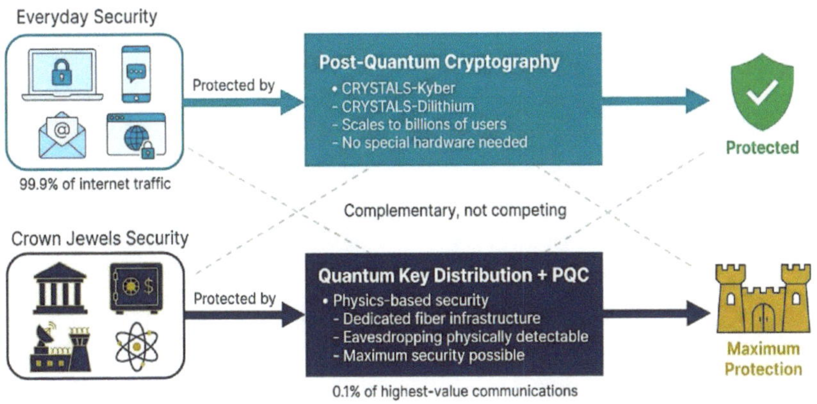

Think of it as home security: Post-quantum cryptography is like having strong locks and an alarm system for your front door, while QKD is like having a dedicated, high-security vault and armed guards for your most valuable possessions.

Think of it like home security. Post-quantum cryptography is your deadbolt—very hard to pick, good enough for most threats. QKD is a vault with armed guards and laser tripwires—overkill for daily life, essential for crown jewels.

Most of us will use post-quantum cryptography and never think about it. A small number of governments and organizations protecting genuinely critical secrets will use QKD. Both will coexist. Both serve different needs.

The internet doesn't run on one security model. It runs on layers of security matched to threat levels and resource constraints. Quantum technology just adds new options to both ends of that spectrum.

What This Means for You

"Should I worry about quantum computers stealing my passwords?"

Probably not.

Timeline: Fault-tolerant quantum computers capable of running Shor's algorithm are at minimum a decade away, possibly two. Your passwords will change dozens of times before then.

Targets: Adversaries capable of "harvest now, decrypt later" attacks are targeting diplomatic communications, military secrets, corporate

IP worth millions, long-term state secrets. Not your Netflix password. Not your email. The economic threshold for what's worth attacking stays high even when decryption becomes easier.

Automatic protection: When you update your browser, operating system, or messaging apps, you're already getting post-quantum cryptography. Chrome deployed it in 2023—if you updated to version 116 or later, you automatically enabled hybrid post-quantum key exchange for all HTTPS connections.[22] You didn›t need to do anything. The protection arrived in a routine update. iMessage did it in 2024.[23] Signal implemented it. The transition happens silently in the background while you sleep.

But use some common sense. Strong passwords still matter for defending against classical attacks, which will remain the dominant threat for years. Two-factor authentication doesn't rely on encryption at all and works regardless of who has quantum computers. Keep your software updated—those routine updates are delivering post-quantum protection whether you notice or not.

And don't panic. For once in cybersecurity, we're ahead of the threat instead of scrambling after it arrives.

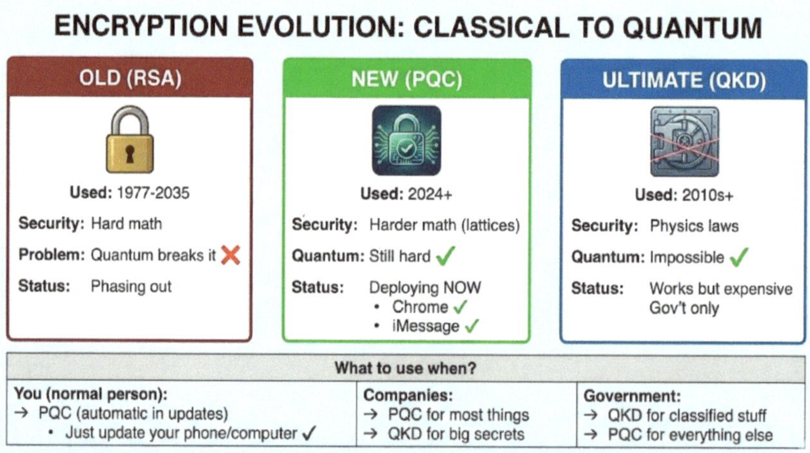

Three encryption types: RSA (old, quantum-vulnerable by 2030s), Post-Quantum Cryptography (new NIST standard 2024, quantum-resistant, deploying in Chrome/iMessage), Quantum Key Distribution (physics-based, unhackable but expensive, government/banks only). Recommendation: Keep devices updated for automatic PQC.

For fifty years, we've trusted mathematics to protect our secrets. Prime factorization is hard. Discrete logarithms are hard. Cryptographic hash functions are hard.

And they are hard—for classical computers.

Quantum computers change the rules completely. Shor's algorithm turns "impossibly hard" into "solvable in hours." The encryption protecting trillions of dollars in transactions, classified military plans, and private messages becomes suddenly vulnerable.

But the same quantum mechanics that threatens security also provides the solution. Post-quantum cryptography uses mathematical problems that even quantum computers struggle with—problems without the periodic structure that quantum algorithms exploit. Quantum Key Distribution uses the laws of physics themselves. You can't eavesdrop without being detected because measurement disturbs quantum states. That's not a design choice. That's physics.

The future of security is a strange hybrid: clever mathematics resistant to quantum attacks, combined with quantum physics that makes eavesdropping physically impossible.

We're in a race. Cryptographers building post-quantum defenses versus engineers building quantum computers powerful enough to attack. Governments mandating upgrades versus adversaries harvesting encrypted traffic for future decryption.

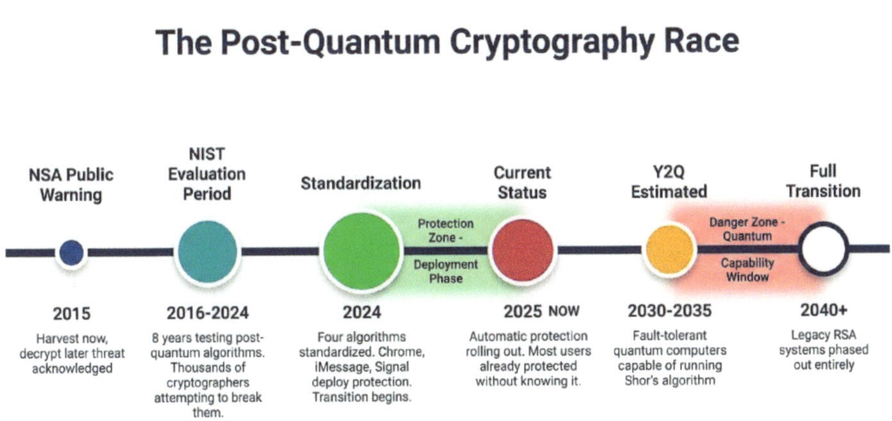

For once in cybersecurity history, the defense arrived before the attack. The upgrades are happening now.

Right now, the defenders are winning. The standardization happened before the attack materialized. The deployment is happening before the threat arrives. For once in the endless cat-and-mouse game of cryptography, the good guys got ahead.

But it's close. Closer than most people realize.

The encryption protecting your secrets today will be obsolete within a decade. The encryption protecting your secrets tomorrow is already here, quietly upgrading in the background while you sleep.

That's the quantum security shift: from mathematical complexity to quantum physics. From trusting that factoring is hard to trusting that measurement disturbs quantum states.

From numbers to nature.

Turn the page.

References

[1] Durumeric, Z., et al. (2013). Analysis of the HTTPS certificate ecosystem. *Proceedings of the 13th ACM Internet Measurement Conference,* 291–304.

[2] Rivest, R. L., Shamir, A., & Adleman, L. (1978). A method for obtaining digital signatures and public-key cryptosystems. *Communications of the ACM,* 21(2), 120–126.

[3] NIST (2024). Key management guidelines. *NIST Special Publication,* 800–57.

[4] Shor, P. W. (1994). Algorithms for quantum computation: discrete logarithms and factoring. *Proceedings of the 35th Annual Symposium on Foundations of Computer Science,* 124–134.

[5] Lenstra, A. K., et al. (2001). Selecting cryptographic key sizes. *Journal of Cryptology,* 14(4), 255–293.

[6] Roetteler, M., Naehrig, M., Svore, K. M., & Lauter, K. (2017). Quantum resource estimates for computing elliptic curve discrete logarithms. *ASIACRYPT 2017,* 241–270.

[7] Mosca, M. (2018). Cybersecurity in an era with quantum computers: will we be ready? *IEEE Security & Privacy,* 16(5), 38–41.

[8] National Security Agency (2024). Quantum computing and post-quantum cryptography FAQ. NSA Cybersecurity Directorate.

[9] Gidney, C., & Ekerå, M. (2021). How to factor 2048-bit RSA integers in 8 hours using 20 million noisy qubits. *Quantum,* 5, 433.

[10] Fowler, A. G., et al. (2012). Surface codes: towards practical large-scale quantum computation. *Physical Review A,* 86, 032324.

[11] Arute, F., et al. (2019). Quantum supremacy using a programmable superconducting processor. *Nature,* 574, 505–510.

[12] Acharya, R., et al. (2024). Quantum error correction below the surface code threshold. Nature, 614, 676–681.

[13] IBM Quantum (2023). IBM Quantum Condor: 1,121-qubit processor. IBM Research Technical Report.

[14] National Security Agency (2015). The next generation of encryption. NSA Suite B Cryptography Announcement.

[15] National Security Agency (2024). Commercial National Security Algorithm Suite 2.0. NSA Cybersecurity Advisory.

[16] Alagic, G., et al. (2024). Status report on the third round of the NIST post-quantum cryptography standardization process. *NIST Interagency Report,* 8413.

[17] NIST (2024). Post-quantum cryptography standardization. Federal Information Processing Standards.

[18] Avanzi, R., et al. (2020). CRYSTALS-Kyber algorithm specifications and supporting documentation. NIST PQC Standardization.

[19] Ducas, L., et al. (2020). CRYSTALS-Dilithium algorithm specifications and supporting documentation. NIST PQC Standardization.

[20] Bernstein, D. J., et al. (2020). SPHINCS+ algorithm specifications and supporting documentation. NIST PQC Standardization.

[21] Fouque, P. A., et al. (2020). FALCON algorithm specifications and supporting documentation. NIST PQC Standardization.

[22] Google Security Blog (2023). Protecting Chrome traffic with hybrid Kyber KEM. Google Security Announcement.

[23] Apple Security (2024). iMessage with PQ3 cryptographic protocol. Apple Platform Security Guide.

[24] Moody, D., et al. (2024). Status report on the third round of the NIST post-quantum cryptography standardization process. *NIST Interagency Report,* 8413-upd1.

[25] Gisin, N., Ribordy, G., Tittel, W., & Zbinden, H. (2002). Quantum cryptography. *Reviews of Modern Physics,* 74(1), 145–195.

[26] Bennett, C. H., & Brassard, G. (1984). Quantum cryptography: public key distribution and coin tossing. *Proceedings of IEEE International Conference on Computers, Systems and Signal Processing,* 175–179.

[27] Liao, S. K., et al. (2017). Satellite-to-ground quantum key distribution. *Nature,* 549, 43–47.

[28] Chen, Y. A., et al. (2021). An integrated space-to-ground quantum communication network over 4,600 kilometers. *Nature,* 589, 214–219.

[29] ID Quantique (2009). First commercial quantum key distribution system deployed for Swiss national elections. Company Press Release.

[30] Korea Telecom (2021). Quantum cryptography communication network in Seoul metropolitan area. KT Corporation Technical Report.

[31] Battelle Memorial Institute (2023). Quantum network deployment for secure government communications. Battelle Technical Report.

[32] Hughes, R. J., et al. (2013). Practical free-space quantum key distribution over 10 km in daylight and at night. *New Journal of Physics,* 4, 43.

[33] European Commission (2023). European Quantum Communication Infrastructure initiative. EU Quantum Flagship Programme.

[34] Sikeridis, D., Kampanakis, P., & Devetsikiotis, M. (2020). Post-quantum authentication in TLS 1.3: a performance study. *Proceedings of the Network and Distributed System Security Symposium.*

[35] UK Quantum Communications Hub (2024). National quantum network phase 1 deployment and performance evaluation. EPSRC Quantum Technology Hub Technical Report.

[36] Nakashima, E. (2023). Intelligence officials warn that adversaries are harvesting encrypted data for future decryption. *Washington Post*, February 14, 2023.

CHAPTER

15

The Quantum Internet

(NOT WHAT YOU THINK)

When Qubits Start Making Long-Distance Calls

What Is the Quantum Internet?

In December 2017, a Chinese satellite transmitted two signals. One to a ground station in Delingha, Tibet. Another to a receiving station in Nanshan, 1,200 kilometers away—roughly the distance from New York to Chicago.

Both signals arrived. Both were measured.

And in that instant, something impossible happened.

The measurement in Tibet changed the measurement in Nanshan. Not through any signal traveling between them—there wasn't time, even at light speed. Not through any physical connection—1,200 kilometers of air, mountains, and the curvature of Earth separated them. The measurements correlated instantly, across a quarter of China, with no mechanism that makes intuitive sense.[1]

> *Einstein called this "spooky action at a distance" and spent the last thirty years of his life trying to prove it couldn't be real.[2] He lost. The universe won. And now China is building an internet on top of it.*

Let me be honest with you about something before we go further. When I first heard the phrase "quantum internet," my brain immediately filled in the blanks: faster streaming, lower latency, finally being able to video call my wife without the thirty-second delay that makes us both sound like astronauts.

None of that. The quantum internet will not make your Netflix faster. I know. I'm disappointed too.

What it does instead is so much stranger—and ultimately so much more important—that the disappointment passes quickly.

The quantum internet isn't the regular internet, but faster. Information still travels at the speed of light—physics doesn't allow otherwise. It won't download your movies quicker. It won't reduce your Zoom lag. Your Netflix stream will remain stubbornly limited by the same old photons trudging through the same old fiber optic cables.

What the quantum internet does is far stranger.

It sends quantum states—entangled particles whose measurements correlate in ways that classical physics cannot explain—between locations in a manner that is fundamentally, physically impossible to intercept without immediate detection. Where classical encryption can theoretically be broken given enough time and computing power, quantum communication cannot be eavesdropped without destroying the message itself.[3]

Think of the classical internet as sending locked boxes through the mail. With enough time, clever mathematics, or a quantum computer running Shor's algorithm, someone can pick the lock and read your secrets without you ever knowing they were there.

> *The quantum internet sends boxes that burst into flames the instant anyone besides the intended recipient touches them. Not metaphorically. Physically.*

The quantum states collapse, the information vanishes, and both sender and receiver immediately know someone tried to intercept.

That's not an incremental security improvement. That's physics replacing mathematics as the foundation of privacy.

And it's being built right now. Not in laboratories as a future possibility. In cities. Between continents. Via satellites orbiting over your head as you read this sentence.

The question isn't whether we'll have a quantum internet. The question is who will control it.

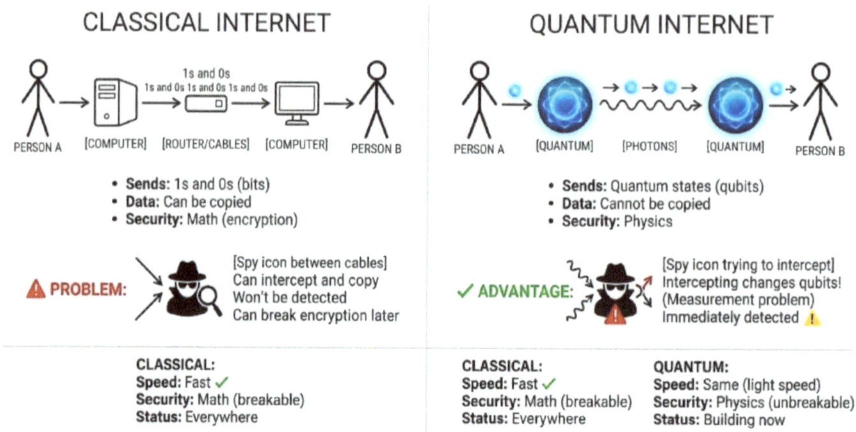

Classical internet sends bits (1s and 0s) secured by mathematics. Quantum internet sends quantum states secured by physics. Interception changes quantum states (measurement problem), making eavesdropping detectable.

Entanglement as Infrastructure

Entanglement showed up in Chapter 5—the phenomenon where measuring one particle instantly determines another particle's state regardless of distance. Einstein hated it. John Bell proved it was real.[4] Alain Aspect demonstrated it experimentally—work that earned the 2022 Nobel Prize in Physics.[5]

What we didn't cover in Chapter 5: using entanglement as infrastructure. As the raw material for a communication network. That requires solving problems nobody thinks about when entanglement is just a curious quantum phenomenon.

Here's something that surprises most people: creating entangled photons is actually not that hard. Shine a laser through a specific crystal—beta barium borate if you want to sound impressive at your next dinner party—and occasionally a single photon splits into two photons[6] that are entangled with each other. Do this a billion times per second and you have a reliable stream of quantum correlations. You can buy commercial entanglement sources for about $50,000—not cheap, but within the budget of most serious research groups. The physics works. We cracked that part.

The hard parts come after creation.

Think of it like this. Generating entangled photons is like making soap bubbles. Easy. A child can do it. The hard part is carrying a soap bubble from your kitchen to your bedroom without it popping. Now imagine trying to carry it from Mumbai to Tokyo. Through air, through glass, past vibrations, through temperature fluctuations—each interaction threatening to collapse the delicate quantum state into ordinary, boring, un-entangled light.

Entangled photons in standard optical fiber survive roughly 100 kilometres before too many of them decohere[7]—the bubble pops, the quantum magic vanishes, you're left with an ordinary photon that's forgotten everything it knew. That 100-kilometre limit is what the entire field of quantum networking is racing to overcome.

Storing entanglement until you need it is the second challenge. You cannot generate entangled pairs on demand precisely when two people want to communicate. Time zones exist. People in Tokyo and London want to connect at arbitrary moments across twelve hours of separation. You need quantum memory to hold entangled states until they are used. Current best quantum memories manage a few seconds for photons, up to minutes for certain atomic quantum states, maybe hours in specialized laboratory conditions.[8] You need days for a practical global network where anyone can connect with anyone else on human timescales.

Distributing entanglement without destroying it is the third. You cannot copy quantum states—the no-cloning theorem from Chapter 12 prevents it. You cannot check if photons are still entangled without measuring them, which destroys the entanglement. You're shipping incredibly fragile quantum correlations across continental distances with no way to verify they arrived intact until the moment you use them. At which point they're gone.

This is why quantum internet is genuinely hard. You're building infrastructure from the most delicate phenomena in physics, with no error-checking available, requiring storage technology that barely exists, across distances where 99% of your signal disappears.

It's like building the classical internet, except every data packet self-destructs if you look at it wrong, you can't verify delivery until the recipient opens it, and 99% of packets vanish in transit anyway.

Except we're doing it. Because the physics makes certain things possible that no amount of classical engineering can achieve.

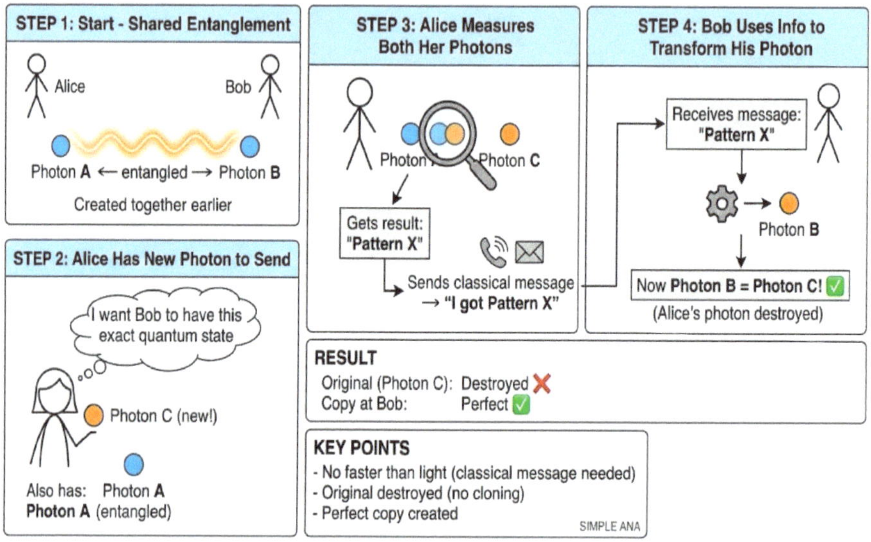

Quantum Teleportation: Terrible Name, Real Physics

The term "quantum teleportation" was coined in 1993 by physicists who apparently wanted maximum public confusion.[9] It doesn't teleport matter. It can't send information faster than light. It requires classical communication to work. It's simultaneously the most precisely accurate name and the most misleading name in quantum physics.

What actually happens:

Alice has a photon in some quantum state—polarized at 37 degrees, or in a specific superposition of vertical and horizontal, or something more complex. She wants Bob to have a photon in exactly the same state. But she can't physically send her photon. Too fragile, too far, doesn't matter why.

In physics, Alice and Bob are the conventional names for the two parties in any quantum communication scenario—a tradition that saved generations of textbook writers from using 'Person A' and 'Person B.

She and Bob previously shared an entangled pair of photons. Created earlier, distributed, stored in quantum memory, waiting for this moment.

Alice performs a Bell state measurement. She takes her photon—the one she wants to send—and her half of the entangled pair, and measures them together in a specific way that entangles them while destroying both. This measurement produces two classical bits of information. Just two ordinary bits. Zeros and ones.

Alice sends those two bits to Bob through completely normal channels. Email works. Phone call works. Carrier pigeon works. No quantum channel needed for this part.

Bob receives the two classical bits. Depending on which two bits Alice sent, he performs one of four possible operations on his half of the entangled pair. After this operation, Bob's photon is now in exactly the same state Alice's original photon was in.

The quantum state has teleported. Alice's photon is destroyed. Bob's photon is now identical to what Alice started with. And at no point did the quantum information travel through the space between them.

The original quantum state vanished in Tibet. The identical quantum state appeared in Nanshan. The information never existed in the 1,200 kilometers between.

This was first demonstrated experimentally in 1997.[10] Then across longer distances through fiber. China's Micius satellite stretched it to 1,400 kilometers from ground station to satellite in 2017.[11] The physics works. We're just getting better at it.

Why does this matter for quantum internet? Classical networks route data packets through intermediate nodes. New York to London might go through Chicago, then Halifax, then across the Atlantic, through multiple routers and switches. The data exists at every point along the path.

Quantum networks can teleport states directly from endpoint to endpoint using entanglement as the quantum channel. The quantum information never exists in between. You can't tap a quantum internet connection in the middle because there's no middle to tap. The quantum state is in Tokyo, then it's in London, and it was never anywhere else.

That's not how any previous communication system has ever worked. And it's the foundation of why quantum internet security is fundamentally different from anything we've built before.

The catch: you need that pre-shared entanglement. Which is exactly the problem quantum repeaters are designed to solve.

The Problem: Physics Eats Your Photons

Standard optical fiber—the same fiber carrying your Netflix stream—loses about 0.2 decibels per kilometer. For classical signals, this is manageable. Amplifiers every 50 kilometers read the signal, boost it, send it forward. The internet crossed oceans using this approach.

For quantum signals, the losses are brutal.

At 50 kilometers, roughly 90% of all photons have been absorbed.

At 100 kilometers, 99% are gone.

At 200 kilometers, 99.99% have vanished.

Beyond 500 kilometers, effectively none survive.

Classical amplifiers cannot solve this. An amplifier must measure the signal to boost it—but measuring a quantum state collapses it. The no-cloning theorem makes it impossible to copy the quantum state before measuring.[12] The amplifier would destroy the very information it's trying to preserve.

Here's the thing that should bother you about this. We solved the distance problem for classical internet in the 1980s with a beautifully simple trick: every fifty kilometres, stick a box that reads the signal and sends it again, louder. Amplify and repeat. The internet crossed oceans.

Quantum internet cannot use that trick. To amplify the signal, you have to read it. To read a quantum state, you destroy it. It's like trying to photocopy a message written in invisible ink—the act of seeing it erases it.

So engineers invented something far more elegant and far more difficult:

Quantum repeaters

A quantum repeater doesn't read and retransmit your signal. It does something called **entanglement swapping**[13], which sounds like a magic trick and essentially is one. Imagine you want to introduce your friend in Delhi to your friend in London, but you can only introduce people who are physically in the same room. So you introduce Delhi-friend to someone in Dubai. Then Dubai-friend introduces themselves to London-friend. And somehow—this is the quantum part—Delhi and London are now connected even though they never met. The middleman creates the relationship without ever carrying a message between them.

The physics is proven. The mathematics works. The engineering required to actually do this across continents in real time is somewhere between "extraordinarily difficult" and "we might need a decade."

The concept: you want to create entanglement between New York and London. But photons can't make it that far. So you place a quantum repeater in Halifax, roughly halfway.

Step 1: Create entangled pairs—one pair shared between New York and Halifax (A–B), another pair shared between Halifax and London (B–C).

Step 2: The Halifax repeater performs a Bell state measurement on both photons it holds—the B photon from the New York pair and the B photon from the London pair. This measurement entangles the A and C photons even though they've never directly interacted and are now separated by 5,000 kilometers.[14]

Step 3: New York and London are now entangled. They can use this entanglement for quantum communication or quantum teleportation without their photons ever traveling the full distance.

Step 4: Chain dozens of repeaters across the Atlantic. Each repeater extends the entanglement another hop. Build a quantum highway from entanglement segments.

The physics is proven. The mathematics works. The challenge is pure engineering, and it's ferocious.

Quantum repeaters need quantum memory to store entangled states. They need near-perfect Bell state measurement operations—

errors compound exponentially across many repeaters. They need synchronization accurate to nanoseconds across continental distances because entanglement swapping only works if the measurements happen within the coherence time.

That synchronization requirement deserves emphasis. If you're swapping entanglement across twenty repeater nodes spanning 10,000 kilometers, every single repeater needs to know when to perform its measurement within a few nanoseconds. GPS provides microsecond timing—a thousand times too slow. You need atomic clocks at every node, fiber-connected time transfer networks, or quantum communication itself to distribute timing. The classical control infrastructure for a quantum network is itself a hard engineering problem.

Current quantum memories hold photonic states for milliseconds. Atomic quantum memories can reach seconds or minutes in laboratory conditions.[8] We need hours to days for a practical network where people in different time zones can establish quantum communication on demand.

The Netherlands' QuTech demonstrated a primitive three-node quantum network in 2021—nodes separated by a few kilometers, but it worked.[15] Deployment-ready quantum repeaters spanning hundreds of kilometers between nodes are probably five to ten years away. Maybe longer. That's the optimistic timeline.

QUANTUM REPEATER NETWORKS: SOLVING THE DISTANCE PROBLEM

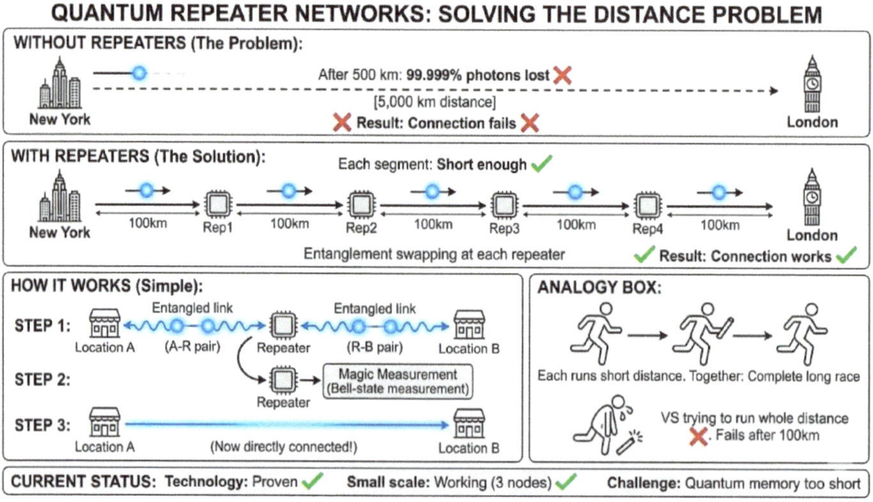

Trusted-Node Networks: The Quantum Internet of 2025

We don't have quantum repeaters at scale yet. But we have trusted-node quantum networks operating right now. Commercial. Operational. Protecting real secrets.

A trusted-node network places measurement stations at regular intervals. Quantum signals travel from node to node, get measured, get re-encrypted using new quantum-generated keys, and are sent to the next node. It's not fully quantum end-to-end—you're trusting each intermediate node not to spy. In China's national network, this means trusting the Chinese state at each node. In any network, the security guarantee is only as strong as the least trusted node—which is precisely why quantum repeaters, not trusted nodes, are the long-term goal. But each link between nodes uses quantum key distribution, which is provably secure: eavesdropping on the key exchange requires measuring quantum states, and measurement leaves a detectable fingerprint in the error rate.[3]

China has built the world's most extensive quantum network—an integrated space-to-ground system spanning 4,600 kilometers connecting Beijing, Shanghai, Jinan, and other cities, with the Micius satellite providing intercontinental reach.[16] It's operational for government communications and financial transactions, verified by European research groups who ran international QKD tests on the network.

Europe is building a continent-wide quantum network—the European Quantum Communication Infrastructure—integrating fiber and satellite links across Germany, Austria, Italy, the Netherlands, and other member states, with a coordinated target of connecting them into a European quantum internet backbone by 2027.

South Korea's Seoul network, already described in Chapter 14, connects government facilities and major banks across 217 kilometres. The United States has research networks at Los Alamos National Laboratory and through Battelle connecting government facilities, and the Illinois Express Quantum Network became operational in 2024—a

52-mile fiber loop connecting Argonne, Fermilab, Northwestern, and the University of Chicago. These are real, working systems. But there is no coordinated national quantum internet programme. No timeline. No unified architecture.

Commercial companies are entering the space. Aliro Quantum sells quantum networking hardware and software. Qunnect provides quantum-secure fiber networks for enterprise customers. These aren't research projects. They're commercial products with paying customers.

The quantum internet of 2025 is like the internet of 1975—functional, proven, limited in scope, expensive, and improving rapidly. The networks work. But they're slow (kilobits per second for key distribution, not the gigabits you're used to) and expensive (hundreds of thousands of dollars per node).[17] It's early. But it's real.

I want to sit with that comparison for a moment. In 1975, the internet was ARPANET—a network connecting a few dozen US universities and military research facilities. The people using it were mostly sending academic papers and playing text-based games. Nobody was running a business on it. Nobody was streaming music. Nobody had predicted Wikipedia, social media, or Zoom.

What they had was the infrastructure. And the infrastructure was enough.

The quantum internet of 2026 is connecting government facilities, university research labs, and a handful of banking networks in China and Europe. It runs at kilobits per second—orders of magnitude slower than your home Wi-Fi. It costs hundreds of thousands of dollars per node.

And it is enough. Because infrastructure always precedes application. The application arrives after someone figures out what this strange new infrastructure makes possible.

What will the quantum internet's equivalent of YouTube be? I genuinely don't know. That's not a cop-out. That's the honest answer—and it's one of the more exciting things about being alive in this particular moment.

Quantum advantage Vs quantum supremacy

In 2019, Google made a headline claim: their 53-qubit Sycamore (Sycamore has 54 qubits; 53 were used in the quantum supremacy demonstration) processor had achieved "quantum supremacy"—it had performed a calculation in 200 seconds that would take the world's best classical supercomputer 10,000 years.

IBM immediately replied: actually, the same calculation would take a classical computer 2.5 days if you used better algorithms and more storage. So no supremacy.

Google said: but we used the hardware we have, and they'd need hardware they don't have.

IBM said: irrelevant, the claim was about speed, not hardware.

Both sides were technically correct. And both sides were missing the point.

Here's the cleaner way to think about it:

Quantum Supremacy = doing *something* a classical computer cannot do in reasonable time. The thing can be completely useless. Google's 2019 task was essentially a random sampling problem with no practical application. It was a demonstration. "We exist. We work. Pay attention."

Quantum Advantage = doing something *useful* that a classical computer cannot do as efficiently. This is the bar that actually matters for drug discovery, financial optimisation, materials science—the applications everyone is actually excited about.

As of 2026: Supremacy—achieved, arguably, multiple times. Advantage—demonstrated in narrow chemistry problems. Broadly transformative advantage—coming, but not here yet.

The reason this distinction matters: if you confuse "supremacy" with "we can now use quantum computers for everything," you'll be disappointed and conclude the hype was wrong. If you understand "advantage is domain-specific and arriving in stages," you'll understand why serious companies are building quantum chemistry teams *now*, before the machines are fully capable. The race is to be ready when advantage arrives—not to wait until it does.

John Preskill, who coined the term NISQ (Noisy Intermediate-Scale Quantum) and also suggested "quantum supremacy," has since expressed regret about the latter phrase. The better term, he now says, is quantum computational advantage. Less dramatic. More accurate. Unfortunately, "supremacy" got to the headlines first.

The Micius Advantage: Why China Is Ahead

In 2016, China launched Micius, the world's first quantum communication satellite—600 kilograms, orbiting at 500 kilometers altitude. Why satellites? Because photons travel through vacuum with almost no loss. No air molecules to scatter them. No fiber impurities to disturb their quantum states. A satellite 500 kilometers above Earth can send quantum signals to ground stations separated by thousands of kilometers on the surface. You bypass the distance problem by going up.

From this single satellite in 2017, the Micius team demonstrated quantum key distribution between Beijing and Vienna over 7,600 kilometers, quantum teleportation from ground station to satellite over 1,400 kilometers,[11] entanglement distribution between ground stations 1,200 kilometers apart,[1] and the first quantum-encrypted intercontinental video call between China and Austria. These aren't proposals or simulations. They're completed experiments, published in *Science* and *Nature*, verified by independent international research teams.

China has since expanded to an integrated space-to-ground network spanning 4,600 kilometers[16] and launched a second quantum satellite in 2023, with plans for a constellation providing global quantum communication coverage by 2027–2028.

The United States has quantum satellite programmes in development and is testing quantum links to the International Space Station. But there is no funded programme to launch a dedicated quantum communication satellite. The earliest realistic date is probably 2028–

2030. Europe has satellite quantum communication experiments but no dedicated launch scheduled.

China is approximately five to seven years ahead in space-based quantum communication. That might not sound like much until you realize that five to seven years in rapidly developing technology is the difference between deployment and development. Between working systems and laboratory tests. Between infrastructure and research.

This isn't just about science. It's geopolitics. Whoever controls quantum communication infrastructure controls the most secure communication channels on Earth. The timeline matters.

Why You Can't Route Quantum Information the Way You Route Classical Data

Classical networks route data by copying packets through intermediate nodes. Quantum information cannot be copied—the no-cloning theorem[12] is a law, not an engineering limitation. Quantum networks therefore require pre-established entanglement paths rather than reactive routing: you build the connection first, then send information through it. The implications—limited bandwidth, path expiry, high latency to establish connections—are fundamental to quantum information, not temporary obstacles engineering will solve.

So how do you route quantum information from Tokyo to London if you can't copy it to send it down multiple possible paths? You can't. Not the way classical networks do.

Quantum routing requires prearranged entanglement paths. Before Tokyo and London can communicate, they need to establish an entangled link—either directly, which doesn't work over long distances, or through a chain of quantum repeaters that swap entanglement from node to node until Tokyo and London are entangled end-to-end.

This is fundamentally different from classical routing. Classical networks are reactive—you send data, routers figure out the path as it travels. Quantum networks are pre-established—you build the entanglement path first, then send quantum information through it.

Every entanglement path has a lifespan measured by quantum memory coherence time—paths expire and must be refreshed. Bandwidth will be shockingly low by classical standards: kilobits per second, not gigabits[17], even with significant engineering progress. And establishing a new entanglement path across multiple repeater nodes may take seconds, versus milliseconds for classical routing.

These aren't temporary limitations. They're fundamental to how quantum information behaves. The quantum internet will never replace the classical internet for bulk data transfer. It will coexist as a parallel infrastructure for applications where quantum properties matter—secure communication, distributed quantum computing, quantum sensor networks.

What You Actually Do With a Quantum Internet

1. **Provably Secure Communication.** The most immediate application—the one driving current deployments. Government communications, military commands, financial transactions: anything requiring absolute secrecy can use quantum key distribution to generate encryption keys secure by the laws of physics, not mathematical assumptions.[3] Already operational in China, Europe, and South Korea. Expanding.
2. **Distributed Quantum Computing.** Connect quantum computers across cities or continents via entangled links. They can share quantum states directly without classical communication bottlenecks—imagine IBM's quantum computer in New York and Google's in Mountain View working collaboratively on the same quantum computation with shared quantum memory. Small-scale demonstrations exist. Practical distributed quantum computing probably needs mature quantum repeaters—five to ten years away.
3. **Quantum Sensor Networks.** Entangle sensors across large distances. They achieve measurement precision beyond what isolated sensors can reach by using quantum correlations to beat the standard quantum limit. Applications include gravitational wave detection using telescope arrays separated by continental distances, precision timekeeping for

fundamental physics tests, and magnetic field measurements for geophysical surveys. Demonstrated in laboratories. Field deployment requires better quantum memories.

4. **Blind Quantum Computing.** Use quantum computers remotely while keeping your data and computation completely private. The quantum computation happens on someone else's hardware, but the input, algorithm, and output remain encrypted in a way that even the quantum computer operator cannot access.[18] Theoretically proven. Early demonstrations exist. Practical deployment requires quantum internet infrastructure that doesn't exist yet.

These applications sound abstract because the enabling infrastructure is still being built. The internet of 1975 connected research institutions for file transfer and email. Nobody predicted YouTube, social media, or streaming services because the applications follow infrastructure, not the other way around. The quantum internet will enable applications we probably haven't imagined yet.

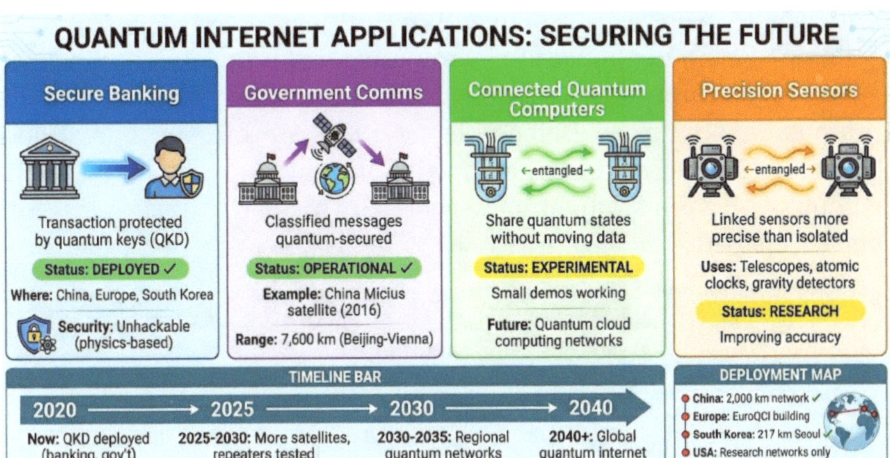

Quantum internet applications: **(1)** Secure banking via QKD—deployed in China, Europe, South Korea. **(2)** Government communications via quantum satellites —China's Micius operational since 2016, 7,600 km range. **(3)** Connected quantum computers—experimental, small demos. **(4)** Precision sensor networks —research phase. **Timeline:** 2025 trus **trusted nodes**, 2030 repeaters, 2040 global network.

Will You Ever Use the Quantum Internet?

Directly? Probably not.

You won't quantum-teleport your vacation photos. You won't entangle your laptop with Netflix servers. Your Zoom calls will remain stubbornly classical, constrained by the speed of light and the bandwidth of your home fiber connection.

What will change—invisibly, behind the scenes—is the security layer protecting certain communications. Your bank will use quantum-generated keys for high-value transactions. Government communications will run over quantum channels. Critical infrastructure—power grids, water systems, financial networks—will have quantum-secured control systems. You won't notice the difference. The padlock icon in your browser might gain a small quantum indicator. That's probably it.

The quantum internet will be infrastructure. Invisible. Boring. Essential. Like electrical grids or sewage systems—nobody thinks about them until they stop working.

The people who will use quantum internet directly: researchers running distributed quantum computing experiments. Intelligence agencies communicating securely. Banks transferring billions between international branches. Defense departments coordinating operations. These users have specific needs—absolute security, quantum computation access, quantum sensor networks—that justify the cost and complexity.

CLASSICAL VS QUANTUM INTERNET

Feature	Classical	Quantum
Signal Type	Light intensity	Entangled photons
Distance Limit	Unlimited (with amplifiers)	~100 km fiber, ~1000 km space (needs quantum repeaters)
Copying	Easy	Impossible (no-cloning theorem)
Eavesdropping	Possible without detection	Physically detectable
Speed	Light speed	Light speed (not faster)
Bandwidth	Gigabits/sec	Kilobits/sec (fundamental limit)
Current Status	Global coverage	Regional networks, expanding

For everyone else, the quantum internet will quietly make certain things more secure while remaining completely invisible. You don't directly use the fiber optic cables under the ocean, but they enable your video calls to Europe. You won't directly use quantum repeaters and entanglement swapping, but they'll enable secure communications that classical physics cannot provide.

The interesting question isn't whether you'll use the quantum internet. It's whether the country you live in will build quantum internet infrastructure or depend on infrastructure controlled by others. China is building it. Europe is building it. The race is on. The United States is falling behind.

The Race That's Already Being Run

The quantum internet is being built right now. Not in some distant theoretical future. In cities across China. In fiber networks across Europe. In satellites orbiting overhead as you read this.

China has operational quantum networks spanning thousands of kilometers,[16] two quantum satellites, plans for a constellation, a national quantum internet programme with billions in funding, and a clear timeline to global deployment.

Europe has regional quantum networks in multiple countries with a coordinated plan to connect them into a continent-wide quantum internet backbone by 2027.

The United States has research testbeds, isolated experiments, and no national programme. No timeline. No coordinated strategy. We invented most of the underlying physics, developed many of the key protocols, and are currently watching other countries build the infrastructure.

This is not a hypothetical competition. It's happening. The question is no longer whether quantum internet will exist. The question is who controls it, who sets the standards, who builds the global infrastructure that will secure communications for the next century.

Thirty years from now, quantum internet might be so ubiquitous that calling it "quantum" will feel antiquated. It'll just be the secure internet—where eavesdropping is physically impossible, where privacy is guaranteed by the laws of nature rather than the complexity of mathematics.

Your grandchildren might take for granted that their communications are absolutely private, that encryption cannot be broken, that data security is a solved problem. When you tell them about the era before quantum internet—when passwords were stolen, when encryption could be broken, when privacy depended on hoping the mathematics was hard enough—they'll look at you the way you looked at your grandparents describing life before the internet existed.

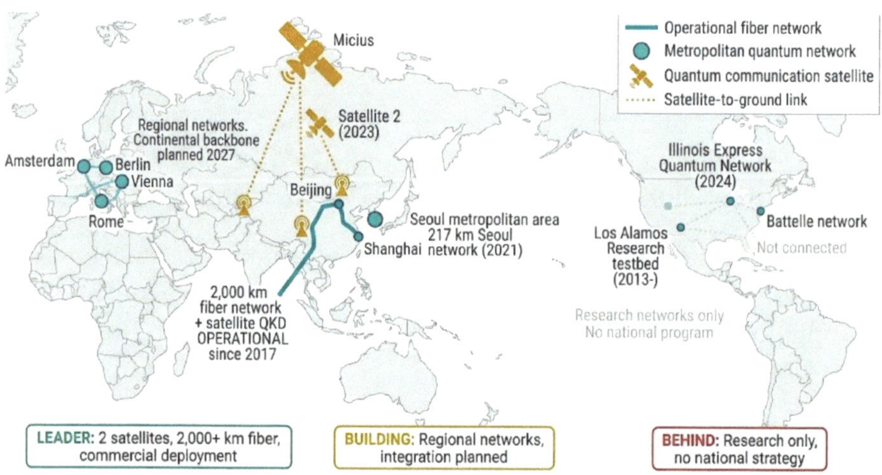

The quantum internet is being built now, not later. Some countries are ahead.

Because to them, the quantum internet won't be the future. It'll be the present. As normal as Wi-Fi. As expected as cellular networks.

But that future requires building infrastructure today. And right now, some countries are building it faster than others.

The race is on. The question is whether you're watching it or running in it.

But there's a problem nobody's talking about yet. A vulnerability that even quantum internet cannot solve. A place where physics-based security meets human-designed systems and creates a gap that might be bigger than anyone realizes.

References

1 Yin, J., et al. (2017). Satellite-based entanglement distribution over 1,200 kilometers. *Science,* 356(6343), 1140–1144.

2 Einstein, A., Podolsky, B., & Rosen, N. (1935). Can quantum-mechanical description of physical reality be considered complete? *Physical Review,* 47(10), 777–780.

3 Gisin, N., Ribordy, G., Tittel, W., & Zbinden, H. (2002). Quantum cryptography. *Reviews of Modern Physics,* 74(1), 145–195.

4 Bell, J. S. (1964). On the Einstein Podolsky Rosen paradox. *Physics Physique Fizika,* 1(3), 195–200.

5 Aspect, A., Dalibard, J., & Roger, G. (1982). Experimental test of Bell's inequalities using time-varying analyzers. *Physical Review Letters,* 49(25), 1804–1807.

6 Kwiat, P. G., et al. (1995). New high-intensity source of polarization-entangled photon pairs. *Physical Review Letters,* 75(24), 4337–4341.

7 Ursin, R., et al. (2007). Entanglement-based quantum communication over 144 km. *Nature Physics,* 3, 481–486.

8 Lvovsky, A. I., Sanders, B. C., & Tittel, W. (2009). Optical quantum memory. *Nature Photonics,* 3(12), 706–714.

9 Bennett, C. H., Brassard, G., Crépeau, C., Jozsa, R., Peres, A., & Wootters, W. K. (1993). Teleporting an unknown quantum state via dual classical and Einstein-Podolsky-Rosen channels. *Physical Review Letters,* 70(13), 1895–1899.

10 Bouwmeester, D., et al. (1997). Experimental quantum teleportation. *Nature,* 390, 575–579.

11 Ren, J. G., et al. (2017). Ground-to-satellite quantum teleportation. *Nature,* 549, 70–73.

12 Wootters, W. K., & Zurek, W. H. (1982). A single quantum cannot be cloned. *Nature,* 299, 802–803.

13 Briegel, H. J., Dür, W., Cirac, J. I., & Zoller, P. (1998). Quantum repeaters: the role of imperfect local operations in quantum communication. *Physical Review Letters,* 81(26), 5932–5935.

14 Żukowski, M., Zeilinger, A., Horne, M. A., & Ekert, A. K. (1993). "Event-ready-detectors" Bell experiment via entanglement swapping. *Physical Review Letters,* 71(26), 4287–4290.

[15] Pompili, M., et al. (2021). Realization of a multinode quantum network of remote solid-state qubits. *Science,* 372(6539), 259–264.

[16] Chen, Y. A., et al. (2021). An integrated space-to-ground quantum communication network over 4,600 kilometers. *Nature,* 589, 214–219.

[17] Kimble, H. J. (2008). The quantum internet. *Nature,* 453, 1023–1030.

[18] Fitzsimons, J. F. (2017). Private quantum computation: an introduction to blind quantum computing and related protocols. *npj Quantum Information,* 3, 23.

CHAPTER

16

Quantum Applications: What Works Now, What's Coming

Drug discovery, battery design, financial optimization, and logistics.

In the summer of 1928, Alexander Fleming left his laboratory in London and went on holiday. He had been growing bacteria cultures—Staphylococcus, the kind that causes boils and pneumonia—and before he left, he made the small, forgivable mistake of leaving one dish uncovered near an open window.

When he came back three weeks later, he found mold growing across the top. Any sensible scientist would have thrown it out. Fleming, for whatever reason—curiosity, boredom, a disinclination to waste a petri dish—looked more closely. The mold was killing the bacteria around it. He wrote up his findings in a 1929 journal paper,[1] received almost no attention for a decade, and eventually helped trigger the discovery of penicillin, which has since saved somewhere between 80 million and 200 million lives, depending on how you count.

Here is the part of that story that nobody puts in the textbooks: we still cannot properly simulate the molecule he found.

> *Penicillin has 41 atoms. Forty-one.*

The electrons in those atoms don't sit in predictable spots like planets in orbit—they smear probabilistically across clouds of possibility, tunneling through energy barriers, forming bonds based on whether their quantum wave functions happen to interfere constructively or destructively. To simulate all of that accurately, the complexity doesn't grow with the number of electrons. It doubles with each one you add.

The world's most powerful supercomputer—including Frontier at Oak Ridge, running at 1.1 exaFLOPS, and El Capitan, which surpassed it in late 2024—can handle molecules of roughly 50 atoms before their approximations start becoming unreliable.[2] Penicillin sits right at that threshold. The drugs that actually matter today are far larger. Remdesivir, the COVID antiviral, has 79 atoms. Tamiflu has 66. Humira, taken daily by millions of people with rheumatoid arthritis, has over 20,000.

This creates a situation that should be somewhat embarrassing for us, given that we can put rovers on Mars and sequence a human genome for $200. When a pharmaceutical company needs to know whether a drug candidate will bind to its target protein, whether it will be toxic, whether it will destroy someone's liver three years after it's approved—they cannot find out from a computer. They have to make the compound, run it through animal studies, then human trials, then wait years. At a cost that industry analyses estimate at $1–2.6 billion per approved drug[3], depending on methodology, with success rates estimated[4] at roughly one in several thousand candidate molecules.

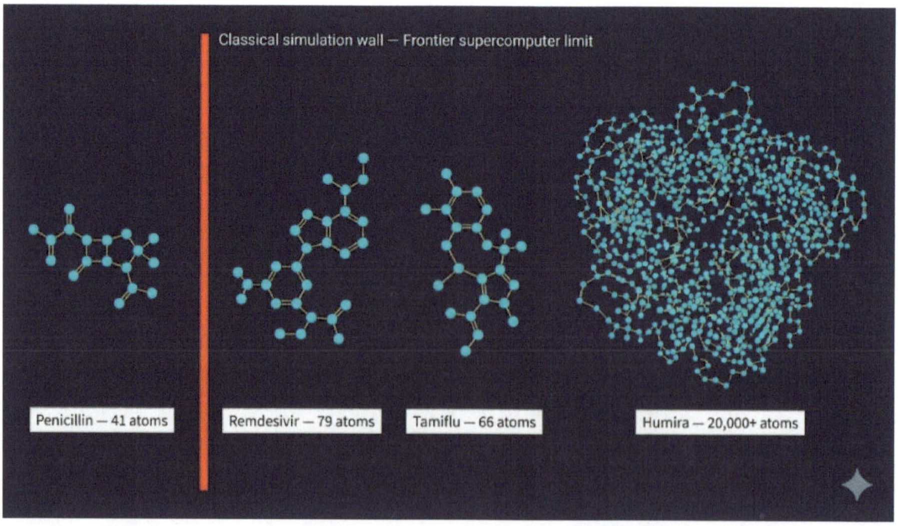

> *Nature runs on quantum mechanics. Classical computers don't speak quantum. And so we approximate—brilliantly, expensively, with genuine scientific heroism—but we approximate nonetheless.*

The Wall Nobody Can Engineer Their Way Around

Let me explain why your computer can't design drugs the same way I'd explain why my grandfather couldn't calculate cricket statistics on an abacus. It's not a skill problem. It's an architectural problem. The tool is wrong for the job.

Here's the fundamental issue. Electrons—the particles that form the chemical bonds in every molecule—don't sit in fixed positions like planets orbiting a sun. Before you measure them, they exist smeared across a cloud of possibilities, simultaneously in many places at once. When two atoms form a bond, their electron clouds interact through quantum interference. The bond forms or doesn't based on whether those wave functions add together or cancel out[5]—a calculation that is *intrinsically quantum mechanical.*

To simulate this on a classical computer, you have to track every possible electron position explicitly. And here is where things get cosmologically absurd.

For N electrons, the number of quantum states you need to follow grows as 2^N.[6] Ten electrons? About a thousand states—manageable. Twenty electrons? A million. Forty? A trillion. And here's the thing about penicillin: the molecule that launched the antibiotic era and has saved somewhere between 80 million and 200 million human lives contains 176 electrons.

2^{176} works out to approximately 10^{53}— a one followed by fifty-three zeros.

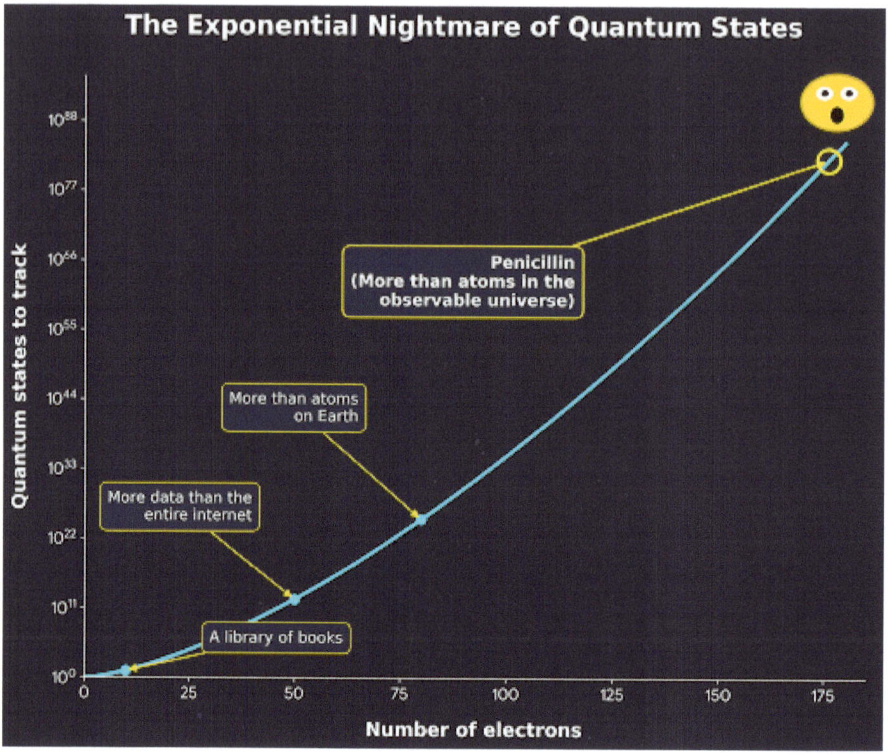

To give you a sense of scale: there are estimated to be around 10^{80} atoms in the observable universe. The calculation required to simulate penicillin's electrons is a number so large it stops being mathematics and becomes philosophy. And penicillin is a *small* molecule. Remdesivir has 79 atoms, hundreds of electrons. Humira, taken daily by millions of people with arthritis, has over 20,000 atoms.

Now you understand why pharmaceutical companies are guessing.

Not approximating. Not estimating. Guessing. Making the compound, testing it in animals, testing it in humans, hoping. At $2.6 billion and fifteen years per approved drug, with a 90% failure rate. The alternative—computing the answer exactly before anyone synthesises anything—requires tools we don't have yet.

Quantum computers are those tools. Eventually.

Chemists, being practical people who have to publish papers and justify salaries, respond to this by simplifying. They drop certain electron interactions, assume some quantum effects are small enough to ignore, use mathematical shortcuts that work well for small molecules and start lying to you around the 50-atom mark. For large molecules, the approximations miss exactly the interactions that determine whether a drug works, whether a catalyst performs, whether a battery holds its charge.[7]

It's why the Haber-Bosch process for making fertilizer—invented in 1913 by a German chemist who later won the Nobel Prize and then helped develop chemical weapons in World War One, making him one of history's more complicated figures—still runs essentially unchanged a century later, burning through 2% of all the energy humanity produces.[8] We can't simulate the catalyst molecules well enough to design something better. Nature solved this problem 3.5 billion years ago using bacteria and an enzyme called nitrogenase. We have studied nitrogenase for decades and still don't fully understand how it works.

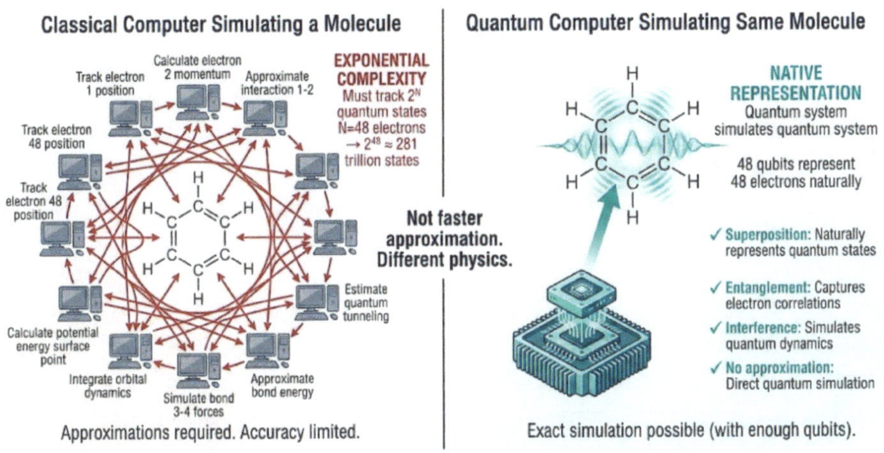

Classical computers approximate quantum behavior with exponentially growing complexity. Quantum computers are quantum systems—they represent molecular behavior natively.

Classical computers aren't going to crack this. The wall isn't made of silicon—it's made of physics.

Feynman's Impatient Proposal

In 1981, Richard Feynman gave a lecture at MIT that opened with roughly the argument above—classical computers can't simulate quantum systems because the state space is astronomical—and then proposed a solution with the impatience of a man who has already figured out the answer and is slightly annoyed at having to explain it.

"Nature isn't classical, dammit," he said, "and if you want to make a simulation of nature, you'd better make it quantum mechanical."[10]

It's an elegant idea, stated with characteristic bluntness. If the thing you're trying to simulate is quantum—if its complexity arises from quantum superposition and interference—then build your simulation machine out of quantum parts. A quantum computer with 176 qubits represents a molecule with 176 electrons not through approximation but natively, because the computer itself is a quantum system governed by the same physics.[9]

You're not asking a classical system to pretend. You're using one quantum system to model another.

Forty-three years after Feynman's lecture, we are finally building the machines he described—noisily, imperfectly, with error rates that would horrify a civil engineer, but building them. The algorithms exist: Variational Quantum Eigensolver runs molecular ground state calculations on current hardware,[11] with others handling energy levels and geometry optimisation. In 2020, Google›s quantum processor ran a simulation of a hydrogen chain—barely a molecule, admittedly, more like a molecular stub—but it did it more accurately than a classical computer could in the same time.[12] In 2023, IBM worked through a 12-electron molecule on 12 qubits. In 2024, Microsoft's Azure Quantum Elements platform let chemists extract useful chemistry results from noisy hardware using error mitigation techniques, and researchers at Pacific Northwest National Laboratory used it for catalytic work that classical approaches couldn't handle.

Small molecules. Incremental progress. Significant caveats about noise and error. But the gap is closing, year by year, in a direction that is consistent if not fast.

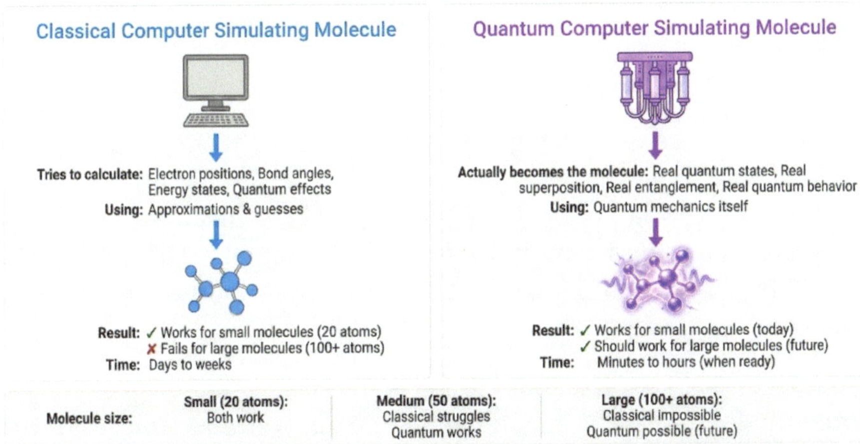

Classical computers approximate molecular behavior using math (works for small molecules, fails for large). Quantum computers simulate molecules using actual quantum mechanics (limited by current hardware size, not fundamental approach). Gap: Need error-corrected qubits for practical drug-sized molecules.

A Lock, A Key, and a Mountain of Wrong Keys

Here's a simple way to understand what quantum computing actually adds to drug discovery, without a single equation.

Your body is full of proteins. Proteins are the machinery of biology—they catalyse reactions, carry signals, defend against infection, build structures. A protein looks, at the molecular scale, a bit like an incredibly complex lock with a three-dimensional shape.

A drug works by being the right key for a specific lock. Slot into the protein in exactly the right way, and you change what the protein does—you block it, or activate it, or prevent it from doing the harmful thing it's been doing.

The problem: there are an almost infinite number of possible keys. And to know if a key works, you need to know not just whether it fits the lock, but *how tightly* it fits, whether it accidentally also fits seventeen other locks in your liver, whether it stays bound long enough to work, and whether it breaks down safely.

That requires knowing the exact quantum mechanical behaviour of every electron in both the key and the lock. AlphaFold (Google DeepMind's protein structure AI) was a revolution because it told us the shape of the lock. Extraordinary. Genuinely Nobel-worthy. But shape alone isn't enough—you need to know the binding energy, the reaction pathway, the quantum chemistry.

That's what quantum computers will eventually provide: not just the shape of the lock, but a precise, quantum-mechanically-accurate simulation of what happens when key meets lock. Screening ten thousand candidate molecules computationally—before anyone synthesises a single one—and advancing only the ones the quantum simulation says will work.

This is why Merck, Pfizer, and Roche are building quantum chemistry teams now, before the machines are ready. Because the team that understands quantum drug simulation when the hardware matures will be five years ahead of the team that starts learning when it arrives.

The Molecule That Might Have Saved Your Grandmother

In January 2020, Moderna's scientists designed a COVID-19 vaccine in 48 hours. Not the manufacturing, not the trials—just the design. The mRNA sequence. Forty-eight hours from the moment the virus's genome was published online to a working molecular blueprint.

It's a remarkable story, though it's worth understanding precisely what it does and doesn't tell us. mRNA vaccines work with a linear sequence—which is fundamentally a code problem, and classical computers handle code problems beautifully. What they can't handle is proteins.

Proteins fold into three-dimensional shapes, and those shapes determine nearly everything in biology: whether a drug binds to its target receptor, whether an antibody neutralises a pathogen, whether an enzyme catalyses the right reaction. Predicting how a protein folds from its amino acid sequence was, for fifty years, one of the hardest open problems in computational biology—a problem so difficult that a dedicated scientific community held annual blind challenges just to track incremental progress.[13]

Then in 2020, DeepMind's AlphaFold 2 solved the structure prediction problem with an accuracy that genuinely shocked the field, including the people who built it.[13] AlphaFold 3, in 2024, went further—predicting not just how proteins fold but how they interact with drugs, DNA, and each other.[14]

This is extraordinary. It is also only part of the problem.

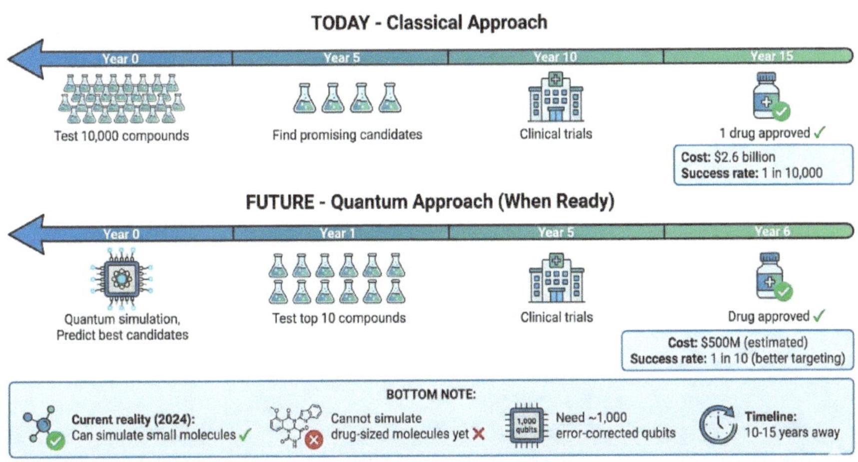

Drug discovery timeline: Classical approach tests 10,000 compounds over 15 years, costs $2.6B, 1 in 10,000 succeed. Quantum approach (future) could simulate best candidates first, reducing timeline to 5-6 years and cost to ~$500M. Current limitation: Need error-corrected quantum computers with 1,000+ logical qubits (10-15 years away).

What AlphaFold doesn't tell you is the binding energy—how tightly a drug molecule actually grips the protein receptor, and whether it grips strongly enough to work. It doesn't model reaction pathways, or predict what happens when your drug also binds to fifteen other proteins in the liver that weren't the intended target. Those questions require quantum mechanical simulation of actual chemistry.

AlphaFold gives you the shape of the lock. Quantum chemistry tells you whether your key fits.

Moderna and IBM have been exploring quantum algorithms for personalised mRNA vaccine design—tailoring immune responses to individual patient genetics, simulating how an mRNA-produced protein will interact with a specific patient's immune receptors. This isn't possible yet. The hardware isn't close. But Moderna doesn't make speculative research bets without running the numbers, and the potential upside—cancer immunotherapies designed to match individual tumour genetics, vaccines personalised to specific population immune profiles—is substantial enough to justify preparation years in advance.

Bristol Myers Squibb, Roche, and Pfizer have all developed quantum computing programmes for drug discovery. Pharmaceutical companies rank among the most conservative investors on Earth. When several of them independently decide that building quantum chemistry teams years before the hardware is ready is the sensible strategy, the implied claim is specific: if you can screen candidates computationally at quantum mechanical accuracy before synthesis, you transform the economics of drug development at every stage.

Industry analysts and quantum chemistry researchers broadly expect useful small-molecule simulation by 2028–2030, with transformative impact[15] dependent on fault-tolerant hardware that most estimates place in the 2033–2037 window. Soon enough that hiring quantum chemists today, before the machines are ready, is the obvious move.

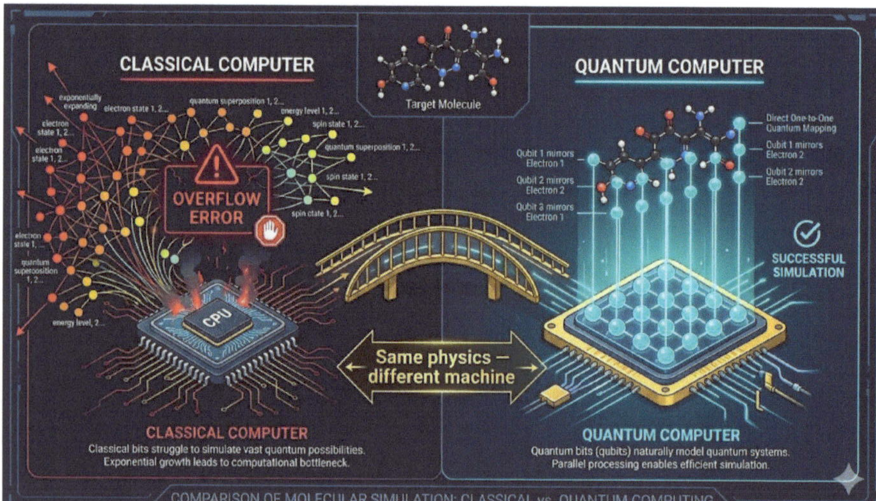

The Battery That Would End an Era

I want to tell you about a battery that exists on paper, is theoretically possible, and would completely transform how humans live on this planet. A battery with 500-mile range. Ten-minute charging. A twenty-year lifespan. Half the cost of what's in your current electric car.

Lithium-sulfur chemistry could theoretically deliver this.[16] So could solid-state batteries. The performance numbers are real. The chemistry is not science fiction.

But here's the problem: we cannot simulate what happens inside these batteries at the molecular level accurately enough to actually build them. Every promising material that looks great in approximation fails in the lab for reasons that our classical simulation didn't catch. Battery engineers call this the "simulation-to-lab gap." I call it "why my electric car still takes forty minutes to charge."

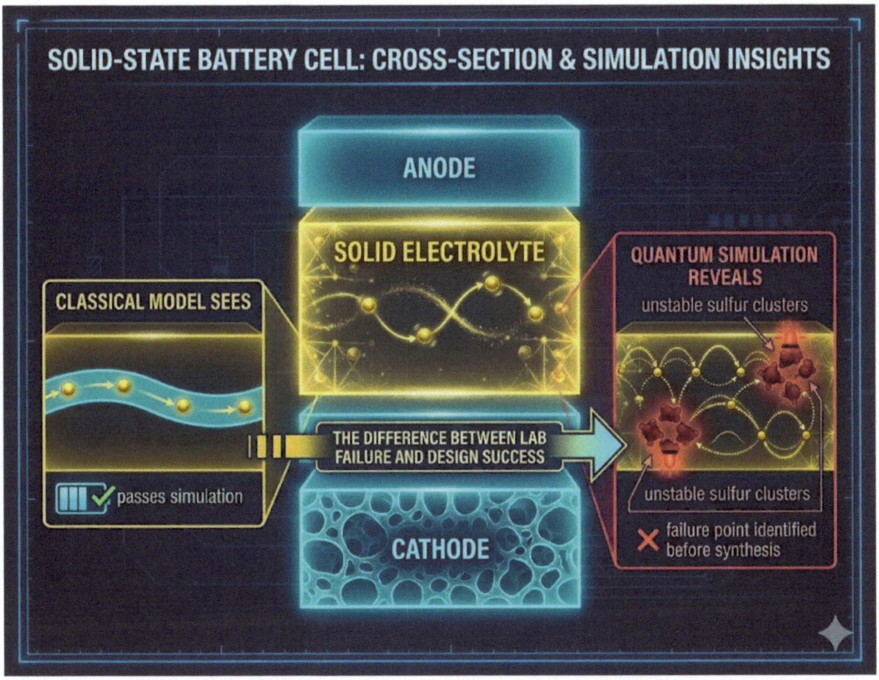

The ions that move through a solid electrolyte quantum-tunnel through energy barriers they classically shouldn't cross. The electrons hop between oxidation states in ways that classical simulation approximates but consistently gets wrong at the detail level that determines whether a battery lasts one thousand charge cycles or five hundred.

Toyota has a multi-billion-dollar bet that quantum simulation will finally crack this. Their commercial target is a solid-state battery in production by 2027–2028. Whether they make that timeline depends on hardware progress that is still uncertain.

But when Toyota—a company so conservative that they spent decades perfecting the petrol engine before touching anything else—makes a multi-billion-dollar quantum computing bet, you pay attention.

Mercedes-Benz has been simulating lithium-sulfur batteries on IBM's quantum systems. The appeal is obvious: lithium-sulfur stores five times the energy per kilogram of lithium-ion in theory; in practice, the batteries fail quickly because the sulfur chemistry degrades in ways that classical models can't fully account for. Quantum simulation could identify exactly how the degradation happens at the molecular level and which electrolyte structures prevent it.

The battery market runs at $120 billion annually today, heading toward $400 billion by 2030.[17] A 10% improvement in energy density saves roughly $5 billion per year in global logistics and energy costs. A 50% improvement makes commercial electric aviation viable. A 100% improvement opens electric cargo shipping—international maritime shipping alone accounts for roughly 3% of global greenhouse gas emissions (International Maritime Organization, 2020) and has no viable electrification pathway on current battery energy density. These aren't incremental improvements to an existing industry. They are the difference between an energy transition that happens this generation and one that doesn't.

The 2% Problem (Which Should Make You Furious)

Here is a fact that should infuriate you.

Somewhere between 1.5% and 2% of all the energy humanity produces goes toward making fertilizer.[18] Not farming. Not irrigation. Not running tractors. Just the one chemical step of turning atmospheric nitrogen—the N_2 that makes up 78% of the air you're breathing right now—into a form that plants can absorb.

This is the Haber-Bosch process. It's been running since 1913. It requires temperatures of 500°C and pressures 200 times higher than normal atmosphere, because nitrogen's triple bond is one of the most stubborn chemical bonds in nature, and the only way we've found to break it at scale is to hit it very, very hard.

The process feeds an estimated 40–50% of the world's population. Without it, roughly half the people alive today would not exist—or would be competing for food that doesn't exist. Fritz Haber, who developed it, won the Nobel Prize in Chemistry in 1918. He also developed poison gas for World War One, making him one of history's most morally complicated figures. But that's a different book.

Here is what makes the fertilizer story interesting for our purposes:

The soil beneath your feet is full of bacteria that do the exact same job.[19] At room temperature. With no external energy input. Using an enzyme called nitrogenase that has been fixing atmospheric nitrogen for 3.5 billion years. We've known about nitrogenase for decades. We've studied it extensively. We still don't fully understand how it works—because the active site involves quantum tunneling and electron spin states that classical simulation cannot accurately model.

We're burning 2% of global energy to do something that bacteria do for free, and the only reason we haven't found a better way is that we can't simulate the enzyme accurately enough to copy it.

BASF—a 160-year-old German chemical company that survived two world wars and is constitutionally allergic to hype—is running quantum simulations of nitrogenase chemistry on IBM's quantum systems. When BASF takes a speculative technology seriously, it means the prize is real.

Cracking nitrogenase computationally could cut global CO_2 emissions by 200 million tonnes annually[20], while still feeding the same number of people. That's not a rounding error. That's more than the aviation industry's entire annual footprint.

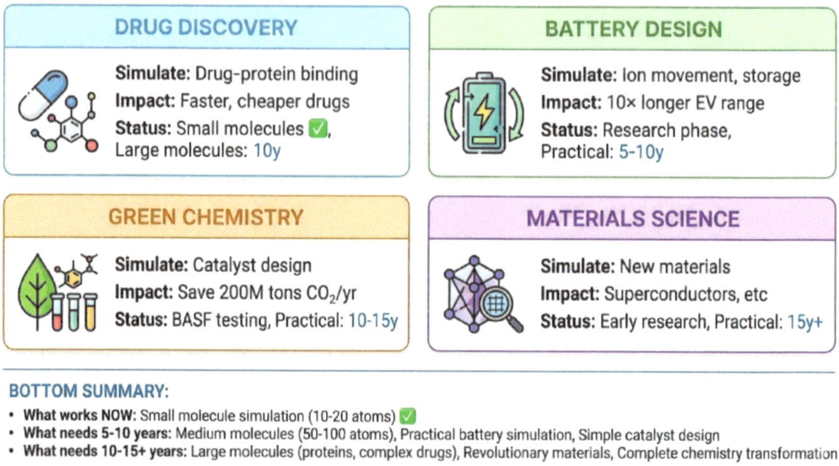

Fusion, Climate, and Other Moonshots

Two other domains deserve mention, with the caveat that honest timelines here require considerably more humility.

Nuclear fusion has famously been twenty years away for about sixty years, which is either a tribute to human optimism or a warning about physics. One genuine computational challenge is plasma simulation: predicting how superheated plasma behaves under fusion conditions involves quantum mechanical effects at small scales—particle collisions, energy transitions, radiation emission—layered on top of macroscopic turbulence and magnetic field dynamics. Current simulations approximate the quantum parts; the approximations limit how accurately fusion reactor performance can be predicted. Several fusion companies are exploring quantum computing for plasma simulation. Whether this will meaningfully accelerate fusion timelines is genuinely unclear. The physics problem is real; the quantum advantage claim is speculative.

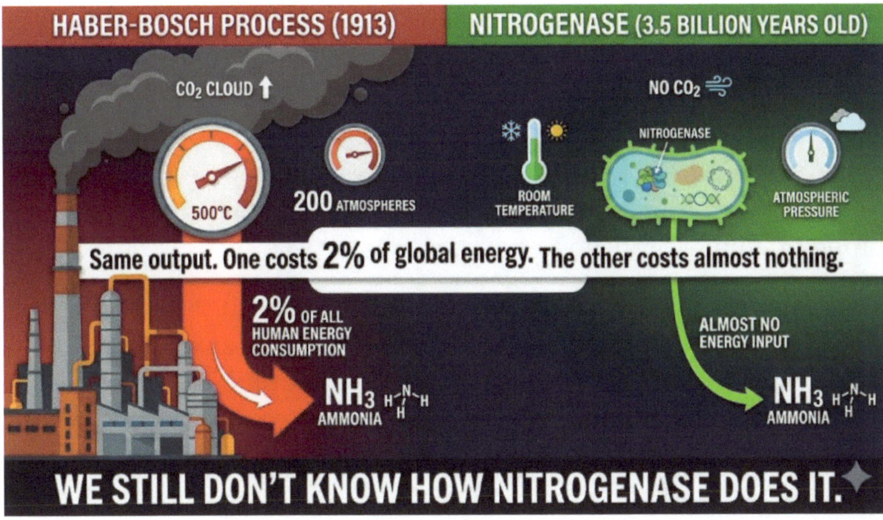

Climate modelling is different. Earth's atmosphere is a quantum mechanical system at the molecular level in ways that actually matter for predictions: greenhouse gas absorption is determined by the vibrational modes of H_2O and CO_2 molecules, cloud formation involves quantum dynamics in aerosol chemistry, photosynthesis and ocean carbon absorption are quantum processes. Current climate models approximate some of this, and improving the quantum mechanical accuracy of greenhouse gas absorption and aerosol chemistry could reduce some of the uncertainty in climate projections—though the dominant sources of uncertainty in climate models are emissions scenarios and large-scale dynamical processes, not molecular-level quantum effects. Quantum computing might reduce some of that uncertainty. How much, and over what timescale, is an active research question with no settled answer.

The Honest Reality Check

Let me save you from a future argument at a family dinner.

Someone—probably the person who just discovered quantum computing via a YouTube thumbnail—is going to tell you that quantum computers will make everything faster. Downloads, AI, games, your laptop, everything. They'll wave their hands and say "quantum" the way people in the early 2000s said "nanotechnology"—as a magical adjective that makes any noun sound futuristic.

They are wrong. Here's why.

Quantum computers are not faster classical computers. They are a fundamentally different kind of computer, suited to a completely different kind of problem[21].

Your laptop runs Microsoft Word. It runs Excel. It runs Chrome. It processes video, plays music, renders web pages, manages your email. All of these tasks involve sequential logic—doing one thing, then another thing, then another. Classical computers are extraordinarily good at this. Quantum computers are not better at it. They are, in fact, worse—slower, more fragile, and requiring cooling equipment the size of a small car.

What quantum computers are spectacularly good at is exploring enormous possibility spaces simultaneously—finding the optimal folding of a protein among billions of possibilities, factoring enormous numbers, simulating quantum chemistry that classical physics cannot model. These are not better versions of the tasks your laptop does. They are completely different tasks.

John Preskill, who coined the term NISQ (Noisy Intermediate-Scale Quantum), compared quantum processors to GPUs. GPUs were designed for video games—parallel graphics processing. When AI researchers discovered in the 2010s that neural network training was also a massively parallel problem, GPUs became the engines of the AI revolution. NVIDIA's market value crossed $4.5 trillion in early 2026. Not because GPUs replaced CPUs—your laptop still has a CPU doing most of the work—but because GPUs were spectacularly suited to a specific class of problem that turned out to matter enormously[22].

Quantum processors will follow the same pattern. Spectacular for quantum simulation and certain cryptographic problems. Irrelevant for everything else. Your laptop stays your laptop. The quantum computer is a specialised instrument—expensive, cloud-accessed, consulted for problems that nothing else can solve.

It will not make your downloads faster. Sorry.

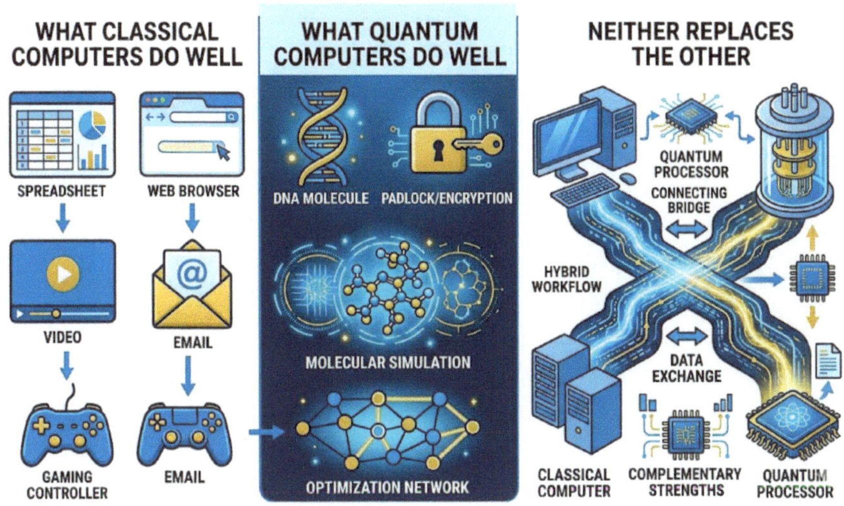

What's Actually Working in 2024

IBM's late-2024 quantum utility demonstration was a genuine inflection point.[23] For specific quantum chemistry simulations—problems where classical approximations start failing—a quantum computer produced results that were useful, verifiable, and better than what classical machines could efficiently generate. The error rates were still higher than production software can tolerate. But "useful results from noisy hardware, without full error correction" is a substantially different category of achievement than the earlier quantum supremacy demonstrations, which were technically impressive and practically useless.

This is the NISQ era finally doing what it promised.[22]

Current working systems: IonQ's Forte Enterprise runs 32 high-fidelity trapped-ion qubits on commercial contracts with pharmaceutical companies—paid work, not pilots. Quantinuum's H2 system achieves gate fidelities above 99.9%, enabling circuits deep enough for real materials simulation, currently running production work for Honeywell's specialty chemicals division. Atom Computing demonstrated a 1,180-qubit neutral atom system, establishing that scaling to thousands of qubits is a solvable engineering problem rather than a physics barrier. Microsoft's Azure Quantum Elements is a production platform that companies are using today to screen molecular candidates before committing to synthesis.

What still doesn't work: simulating molecules larger than roughly 50 atoms with high accuracy. Running Shor's algorithm on cryptographically relevant key sizes. Fault-tolerant quantum computing with full error correction. Large-scale optimisation. The distance from current capability to industry-transforming capability remains significant. But it's a measured distance now, not a frontier that dissolves when approached.

Error Mitigation: The Trick That Bought Us a Decade

For most of the past decade, the standard answer to "when will quantum computers be useful" was: after full error correction. Error correction requires thousands of physical qubits per logical qubit, which meant millions of physical qubits before anything interesting was accessible. That pushed useful applications to 2035 or later.

Then researchers developed error mitigation, which sidesteps the problem rather than solving it.[24] Instead of eliminating quantum errors, you characterise the noise pattern carefully and correct for it statistically—run the same circuit many times with calibrated variations, map the noise, subtract it from the result. The analogy that circulates is noise-cancelling headphones: you don't remove the airplane engine, you model its frequency signature well enough to cancel it from the audio signal. IBM used this approach on 127-qubit systems in 2024 without any error correction and produced genuinely useful chemistry results. Google reported similar outcomes.

Error mitigation isn't a permanent solution. For deep circuits—the kind needed for Shor's algorithm, complete protein simulation, large-scale optimisation—you eventually need fault tolerance. But for the class of quantum chemistry problems tractable on current NISQ hardware, error mitigation has advanced practical timelines by roughly five years. The 2028–2030 window for practical small-molecule quantum chemistry reflects what current trajectories actually show, not optimistic extrapolation.

The Hardware Is Not Stopping

Google's Willow processor in late 2024 demonstrated error correction below the surface code threshold—the critical fault rate below which adding more physical qubits actually reduces total system error rather than adding to it.[25] This is the foundational proof that fault-tolerant quantum computing is physically reachable. IBM's roadmap targets 100,000 qubits by 2033.[26] Eight years ago IBM had 5 qubits. The trajectory is not linear.

Whether specific roadmap milestones translate into practical advantage on economically meaningful problems at the stated timescales is genuinely uncertain—this field has a history of slippage, and roadmaps function partly as investor communications. But the physics is working, the error correction is working, and the gap between laboratory demonstration and commercial utility is narrowing in ways that weren't visible five years ago.

Merck's Nine-Figure Bet

In 2021, Merck entered a commercial partnership with IBM targeting quantum computing for drug discovery.[27] Industry analysts put the multi-year investment in the $50–100 million range. A company that had survived since 1668—through plagues, world wars, and the occasional drug recall—decided to bet nine figures on hardware that at the time could barely simulate a four-atom molecule.

The logic requires understanding the economics of pharmaceutical development, which are genuinely brutal. Developing a new drug costs an average of $2.6 billion from discovery to approval.[3] That number has been rising roughly 10% per year. Clinical trial failure rates are increasing—more candidates fail expensive late-stage trials. Patent protections expire on blockbuster drugs, and the pipeline to replace them is expensive and unreliable. A pharmaceutical company that cannot materially improve its hit rate in early discovery is in structural financial trouble, not immediately but inevitably.

Quantum simulation offers a specific escape route: predict drug-target binding computationally with quantum mechanical accuracy before anything is synthesised. Screen ten thousand candidates computationally, advance only the thousand that the simulation says are promising, run clinical trials on a smaller, pre-vetted pool. Compress a 15-year development timeline toward 7 years. Cut the $2.6 billion average toward something a fraction of that size. Even a 20% improvement in success rates across Merck's entire portfolio—not transformative, just 20% better—would save billions annually.

Merck now has quantum chemists on staff. They have modified their drug development pipeline to incorporate quantum simulation results as the hardware becomes capable of generating them. They are building institutional capability for a technology that doesn't fully exist yet, on the theory that companies that wait until the technology is proven will be five years behind companies that prepared in advance.

Toyota is doing the same thing for battery design. BASF for industrial catalysts. Goldman Sachs for portfolio optimisation. The pattern is consistent across industries with different problems but similar economic logic: the cost of being early is the cost of preparation; the cost of being late may be existential.

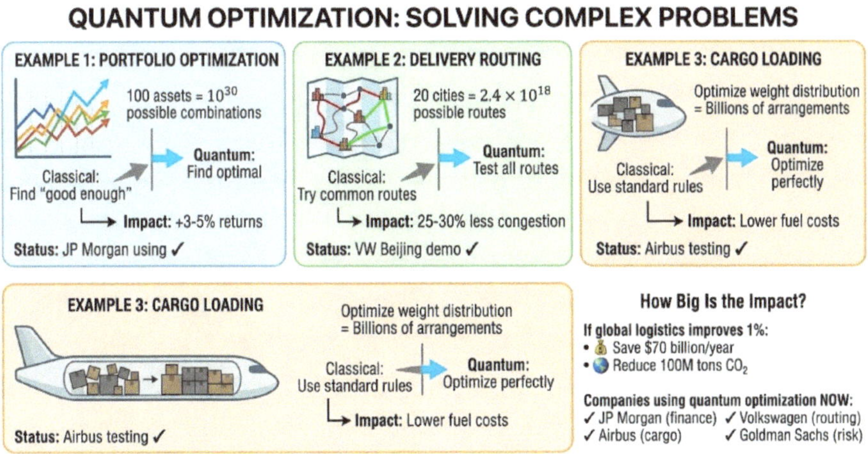

Quantum optimization applications: Portfolio (100 assets = 10^{30} combinations, JP Morgan deployed), routing (20 cities = 2.4×10^{18} routes, VW Beijing demo reduced congestion 25-30%), cargo loading (billions of arrangements, Airbus testing). Impact: 1% global logistics improvement = $70B saved, 100M tons CO_2 reduced annually.

The Molecules Are Waiting

There is a version of this story that ends with a triumphant paragraph about the quantum revolution arriving on schedule and transforming everything. That version would be overstating things.

The honest version is more interesting. The first genuinely useful applications—narrow cases in small-molecule chemistry, specific materials problems, targeted optimisation—are probably coming in 2027–2028. Broader applications by 2030–2032. IBM went from 5 qubits in 2016 with a roadmap to 100,000 by 2033.[26] Google demonstrated error correction below the critical threshold in 2024.[25] These are not press release milestones. They are physics milestones. The transformative applications—complete protein simulation, room-temperature superconductor design, large-scale optimisation—need fault-tolerant machines that won't exist before 2035 at the earliest.

The nature of the uncertainty has changed, though. These are no longer timelines derived from wishful extrapolation. They come from specific companies hitting specific technical milestones on consistent schedules. The molecular simulations that were impossible in 2020 are running under commercial contract today. The error correction thresholds that were theoretical targets in 2022 were demonstrated in hardware in

2024. The gap to transformative capability is real and large. It is also, for the first time, measurable.

The drug that treats Alzheimer's might be designed on a quantum computer. The battery that makes commercial electric aviation viable might be simulated quantum mechanically before a single physical cell is built. The catalyst that makes green fertilizer economical—the one that could end our 110-year dependence on a process invented the year the Titanic set sail—might be found by modelling nitrogenase on hardware that didn't exist five years ago.

There is, however, a constraint that rarely gets talked about in the coverage of quantum computing milestones and investment announcements, partly because it's less photogenic than a new qubit record and partly because admitting it is uncomfortable.

The workforce doesn't exist.

Building a quantum computer is one kind of problem. Programming it—designing and running the chemistry simulations, interpreting results, integrating them into a drug development pipeline or a materials research process—requires a different and rarer kind of expertise. The kind that takes years to develop. The kind that universities are producing in the hundreds annually, while industry is going to need it in the thousands. Possibly tens of thousands.

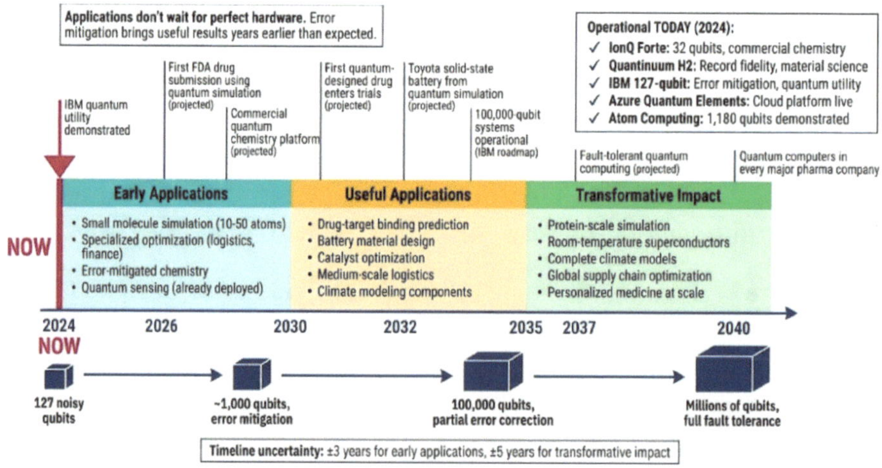

The quantum revolution isn't coming – it's unfolding. Early applications arrive 2026-2028. Transformative impact follows 2035-2040.

The race to build quantum computers has received enormous coverage. The race to train the people who will actually use them has been almost invisible.

Whether we're winning that one is less clear than anyone building the hardware seems comfortable admitting.

References

[1] Fleming, A. (1929). On the antibacterial action of cultures of a penicillium, with special reference to their use in the isolation of B. influenzæ. *British Journal of Experimental Pathology,* 10(3), 226–236.

[2] Oak Ridge National Laboratory (2024). Frontier supercomputer capabilities and molecular simulation benchmarks. ORNL Computing Report.

[3] DiMasi, J. A., Grabowski, H. G., & Hansen, R. W. (2016). Innovation in the pharmaceutical industry: new estimates of R&D costs. *Journal of Health Economics,* 47, 20–33.

[4] Paul, S. M., et al. (2010). How to improve R&D productivity: the pharmaceutical industry's grand challenge. *Nature Reviews Drug Discovery,* 9(3), 203–214.

[5] Szabo, A., & Ostlund, N. S. (1996). *Modern Quantum Chemistry: Introduction to Advanced Electronic Structure Theory.* Dover Publications.

[6] Aspuru-Guzik, A., Dutoi, A. D., Love, P. J., & Head-Gordon, M. (2005). Simulated quantum computation of molecular energies. *Science,* 309(5741), 1704–1707.

[7] Reiher, M., Wiebe, N., Svore, K. M., Wecker, D., & Troyer, M. (2017). Elucidating reaction mechanisms on quantum computers. *Proceedings of the National Academy of Sciences,* 114(29), 7555–7560.

[8] Smil, V. (2004). *Enriching the Earth: Fritz Haber, Carl Bosch, and the Transformation of World Food Production.* MIT Press.

[9] Cao, Y., et al. (2019). Quantum chemistry in the age of quantum computing. *Chemical Reviews,* 119(19), 10856–10915.

[10] Feynman, R. P. (1982). Simulating physics with computers. *International Journal of Theoretical Physics,* 21(6–7), 467–488.

[11] Peruzzo, A., et al. (2014). A variational eigenvalue solver on a photonic quantum processor. *Nature Communications,* 5, 4213.

[12] Google AI Quantum and Collaborators (2020). Hartree-Fock on a superconducting qubit quantum computer. *Science,* 369(6507), 1084–1089.

[13] Jumper, J., et al. (2021). Highly accurate protein structure prediction with AlphaFold. *Nature,* 596, 583–589.

[14] Abramson, J., et al. (2024). Accurate structure prediction of biomolecular interactions with AlphaFold 3. *Nature,* 630, 493–500.

[15] Boston Consulting Group (2024). The emerging quantum computing ecosystem in pharmaceuticals. BCG Technology Report.

[16] Janek, J., & Zeier, W. G. (2016). A solid future for battery development. *Nature Energy,* 1, 16141.

[17] BloombergNEF (2024). Battery Market Outlook 2024. BNEF Energy Storage Analysis.

[18] Smith, C., Hill, A. K., & Torrente-Murciano, L. (2020). Current and future role of Haber-Bosch ammonia in a carbon-free energy landscape. *Energy & Environmental Science,* 13(2), 331–344.

[19] Hoffman, B. M., Lukoyanov, D., Yang, Z. Y., Dean, D. R., & Seefeldt, L. C. (2014). Mechanism of nitrogen fixation by nitrogenase: the next stage. *Chemical Reviews,* 114(8), 4041–4062.

[20] MacFarlane, D. R., et al. (2020). A roadmap to the ammonia economy. *Joule,* 4(6), 1186–1205.

[21] Aaronson, S. (2008). The limits of quantum computers. *Scientific American,* 298(3), 62–69.

[22] Preskill, J. (2018). Quantum computing in the NISQ era and beyond. *Quantum,* 2, 79.

[23] IBM Quantum (2024). Evidence for the utility of quantum computing before fault tolerance. *Nature,* 618, 500–505.

[24] Temme, K., Bravyi, S., & Gambetta, J. M. (2017). Error mitigation for short-depth quantum circuits. *Physical Review Letters,* 119(18), 180509.

[25] Google Quantum AI (2024). Quantum error correction below the surface code threshold. *Nature,* 614, 676–681.

[26] IBM Quantum (2023). IBM Quantum development roadmap. IBM Research Technical Report.

[27] Merck & IBM (2021). Quantum computing collaboration for drug discovery. Joint Press Release.

CHAPTER

17

The Race Nobody's Talking About

And why you should care who wins

The Man Who Went Home

In the late 1990s, a young Chinese physicist named Pan Jianwei arrived at the University of Vienna to study under Anton Zeilinger—at the time the world's leading experimentalist in quantum entanglement.[1] If you were a quantum physicist in the late 1990s and you weren't at Zeilinger's lab, you were somewhere else—and you knew it. Pan had earned his place there. Vienna was comfortable, well-funded, and exactly the kind of city where a promising physicist could spend a productive life, an extraordinary place to build a career.

He didn't stay.

After completing his doctorate in 1999, Pan returned to China and established a quantum photonics laboratory at the University of Science and Technology of China in Hefei in 2001.[2] The conventional path for talented Chinese scientists trained in the West at that time was to stay where the equipment was better, the funding more reliable, and the collaborative networks denser. Pan went home anyway. He had a specific ambition: to build a quantum communication satellite. Not a theoretical paper about quantum communication satellites. An actual one, in orbit, transmitting entangled photons through the vacuum of space to ground stations thousands of kilometres apart.

His mentor in Vienna understood exactly what Pan was trying to do—because Zeilinger had been trying to do the same thing himself.

For years, Zeilinger had been lobbying the European Space Agency to fund a quantum satellite. His group had already demonstrated quantum communication over 144 kilometres between the Canary Islands, and the physics was clearly scalable. ESA's committees weighed the proposal, debated it, and then did what committees do when a project is ambitious and the timeline uncertain. Zeilinger later described the ESA process as so slow that no decision was ultimately made, in comments reported by Nature in 2012.[3]

Pan, watching this from Hefei, drew the obvious conclusion: if the satellite was going to be built, he was going to build it.

He spent the next eight years doing the physics the hard way—ground-based experiments, one distance record at a time, building the experimental case that would eventually convince the Chinese Academy of Sciences to fund him[2]. Formal approval came in 2011. Four years of engineering followed. By 2015 he had a flight-ready instrument. By 2016 Pan had a launch window. Thirteen years from proposal to launch, and the clock was running.

The student who went home – and built what his teacher could not.

Zeilinger, watching from Vienna, did the sensible thing: he joined. In 2013, Austria and China announced the Quantum Experiments at Space Scale project, with a dedicated ground station in Vienna.[3] The student and the teacher would share the experiment—even if the student had built the instrument after the teacher's institution had been too slow to fund one.

1:40 AM, Gobi Desert

On August 16, 2016, at 1:40 in the morning local time, a Long March 2D rocket lifted off from the Jiuquan Satellite Launch Centre in the Gobi Desert.[4] The satellite it carried was named Micius, after an ancient Chinese philosopher who, twenty-five centuries earlier, had conducted the first recorded experiments with light—documenting how images invert when light passes through a pinhole. Pan had chosen the name personally.

> *The launch received almost no coverage in the Western press. To those outside the specialist quantum physics community, it was indistinguishable from the routine cadence of Chinese space missions: another payload, another orbit, another entry in the log.*

What it was not was unremarkable.

Over the following months, Micius did three things, each of which was a world record. It distributed quantum entanglement between ground stations 1,200 kilometres apart,[5] quantum teleportation from a ground station to the satellite 1,400 kilometres above,[6] and a quantum-secured video call between Beijing and Vienna—7,600 kilometres of relay path—that no eavesdropper could intercept without immediate detection.[7]

That call, on September 29, 2017, was between Pan Jianwei in Beijing and Anton Zeilinger in Vienna: teacher and former student, separated by a continent, connected by a satellite the student had built after the teacher's institution had been too slow to act.

The results were published in *Science* and *Nature* in 2017. Pan's team had not submitted one paper. They had submitted three, simultaneously, each reporting a separate world record.

The reaction in Western quantum research communities was a specific variety of professional shock rare in science—the feeling not of being beaten on a problem you were working on, but of discovering that someone else had been working on a different level of the same problem entirely, and had finished, while you were still debating whether it was worth starting.

"Wait," the conversations went. "They actually *built* that?"

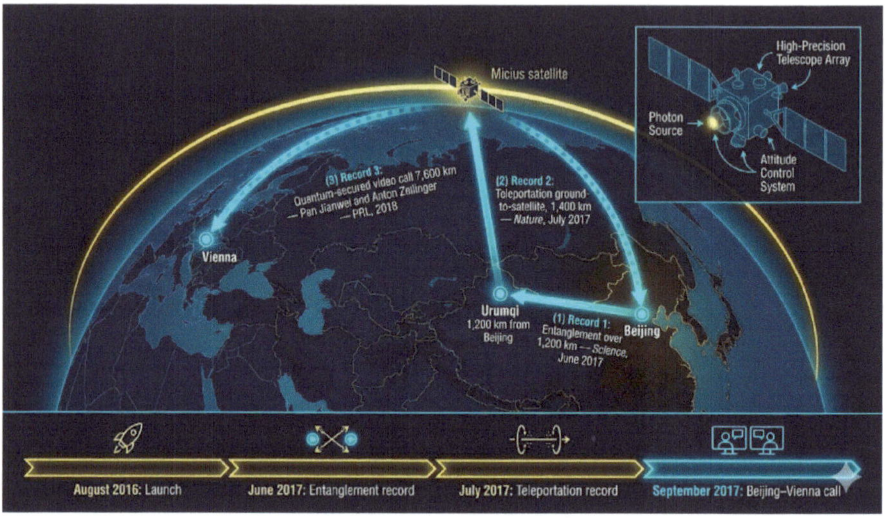

Not in a lab. Not as a proof of principle. As an operational satellite, in orbit, transmitting entangled photons through 1,200 kilometres of open space, reliably, repeatedly, on schedule.

The satellite was roughly the size of a small car. It had taken thirteen years, a national programme, and a physicist who refused to wait for someone else's committee. And it had just changed the strategic calculus of every major nation on Earth.

The $50 Billion Question Nobody Asked Before 2016

The reaction was swift, global, and slightly panicked. Within two years, the United States had passed the National Quantum Initiative Act.[1] The EU expanded its Quantum Flagship programme.[2] The UK, Japan, Canada, India, South Korea, and Australia each announced strategies with serious funding attached.[3–8] The satellite had cost China thirteen years and a national programme. The response cost everyone else their composure.

Here is the uncomfortable arithmetic. China's official quantum investment sits at $15 billion[9]—likely higher once provincial budgets and military programmes are counted. The EU committed €12 billion.[2] The United States matched China's figure through a combination of $3 billion in federal funding and an estimated $15 billion in private sector investment[10]. The UK spent £3.5 billion.[3] Japan (¥75 billion)[4], India (₹8,500 crore)[5], Canada ($500 million)[6], South Korea ($180 million)[7], and Australia ($120 million)[8] added their own programmes.

Global government investment in quantum technologies now exceeds $50 billion.[11] Private capital has added another $12 billion in venture funding since 2018.[12]

GLOBAL QUANTUM TECHNOLOGY INVESTMENT BY NATION

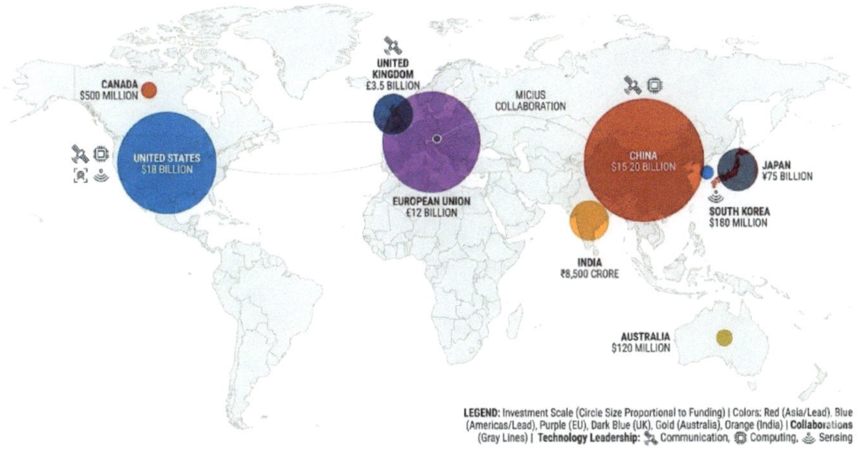

Try to hold that number in your head for a moment. Over sixty billion dollars invested in technology that, in most of its forms, still does not work reliably enough for commercial use. We are talking about hardware that operates at temperatures colder than deep space, collapses in microseconds if you look at it wrong, and is years away from delivering on the promises attached to it. Sixty billion dollars. Spent with intent and eyes open.

Every government that has written one of these cheques knows all of this. They wrote the cheques anyway. Not because they are irrational, but because they have absorbed the lesson that runs through the last two centuries of technological history: the nations that lead a technological revolution do not just win economically. They write the standards, set the terms of international trade, determine who has access to the technology and who does not, and—in ways that are difficult to quantify but impossible to miss—shape the character of the century that follows.

Pan Jianwei, working quietly in Hefei, had made that conclusion unavoidable.

Three Strategies, No Clear Winner

The United States: Innovation Through Chaos

The United States has committed roughly $3 billion federally since the National Quantum Initiative Act passed in 2018, with an estimated $15 billion more from the private sector. The total is impressive. The coordination is not.[10]

The National Quantum Initiative Act, signed into law in December 2018, created a coordinated federal strategy for quantum technology.[1] In practice, "coordinated" means that multiple agencies fund quantum research in largely independent directions.

The NSF funds university research. The Department of Energy runs its own quantum centres. NIST spent eight years on post-quantum cryptography.[13] DARPA pursues applications that, as is DARPA's habit, cannot be publicly discussed. Each agency is doing real work. None of them are talking to each other in any way that could be called a strategy.

It's like five world-class chefs each cooking a different course of a meal, in separate kitchens, with no agreed menu, no shared timeline, and no serving coordinator. The food is excellent. Whether it arrives in any sensible order is another question.

Then there are the technology companies, whose quantum programmes represent something closer to civilisational ambition dressed in corporate language. IBM followed its public roadmap with unusual fidelity—quantum computers exceeding 1,000 qubits, cloud access to thousands of researchers globally[14], a stated target of 100,000 qubits by 2033. Google demonstrated quantum supremacy (disputed) in 2019[15], and then, in 2024, demonstrated with its Willow chip that error correction below the critical fault threshold is physically achievable[16]—the foundational proof that fault-tolerant quantum computing is engineering, not fantasy.

Microsoft is pursuing topological qubits built from exotic quasiparticles, an approach that has had a complicated relationship with experimental verification. A 2018 Nature paper was retracted after data problems were found.[17] 2025 results received cautious editorial notes[18]. If it works, it could produce qubits with inherent error resistance. If it doesn't, it will stand as a cautionary tale about the gap between a promising idea and a working device.

Amazon built the platform connecting customers to everyone else's quantum computers without building its own.[19]

IonQ went public at a $2 billion valuation on $2.1 million in revenue.[20] The valuation reflected expected future capability, not present revenue—a pattern common to technology companies at early commercial stages, and a reasonable bet given the sector's trajectory.[20] Atom Computing demonstrated a 1,180-qubit neutral atom system.[21] Rigetti sells hybrid quantum-classical systems to commercial customers.[22] PsiQuantum has raised over $900 million for a photonic approach that is still in development—which reflects a long-bet architecture thesis: that silicon photonic fabrication at semiconductor scale will ultimately provide the manufacturing pathway to millions of qubits that superconducting approaches cannot easily match. The thesis remains unproven at the required scale. It is also supported by serious technical analysis and backed by governments and major financial institutions who have assessed the risk.[23]

Half a dozen companies building quantum computers that cannot talk to each other, using physics that cannot be compared, targeting markets that do not yet exist, and none of them certain which of their approaches will still be standing in 2035.

It is expensive. It is duplicative. It is also, historically, how America has won technology races—by funding the forest rather than choosing the tree. The Wright Brothers' first flight happened because roughly a dozen teams were attempting powered flight simultaneously and one of them got there first. The internet emerged from ARPANET research that nobody initially intended to commercialise.

The structural weakness is obvious: nobody is driving. IBM and Google compete with each other, not coordinating toward a common goal. Venture capital demands returns on timescales quantum computing cannot yet deliver. The breakthroughs that matter most—fault-tolerant quantum computing, room-temperature operation, large-scale error correction—require patient, decade-scale investment that Silicon Valley is constitutionally unable to provide.

China: Centralised Ambition

China's official quantum investment stands at $15 billion, with the real figure—once you add provincial budgets, military programmes, and what the five-year plans don't advertise—likely closer to $20 billion or above.[9]

While American researchers were debating which approach to fund, China was building the National Laboratory for Quantum Information Sciences— a 37-hectare campus in Hefei, backed by $10 billion in government funding, designed specifically and exclusively to advance quantum technology leadership.[24] Not a university that also does quantum research. Not a corporate lab with a quantum division. A national facility whose entire purpose is to win the quantum race.

What matters more than the money is the time horizon. China's quantum programme operates under central scientific oversight with direct State Council support. It runs on five-year plans that survive political transitions. The scientists running it do not have to justify their existence to venture capital every eighteen months or explain burn rates to public market analysts every quarter. They have a thirty-year roadmap and the institutional patience to execute it.

Think of the difference this way. Running a quantum programme through Silicon Valley is like trying to grow a forest by persuading thousands of individual gardeners that trees are a good investment, with each gardener free to quit when their patience runs out. Running one through Beijing is like the State deciding: we are building a forest, here is the land, here is the budget, here is the twenty-year plan.

Each approach has produced forests. The Silicon Valley forest has stranger trees—unexpected species that nobody planned, mutations that turned out to be valuable. The Beijing forest is taller, more uniform, and built on schedule.

Which forest wins the race? Ask again in 2035.

The results are tangible. Two quantum communication satellites operational. The Beijing-Shanghai quantum network extended to over 4,600 kilometres, carrying government and financial communications.[25] Quantum computing systems exceeding 100 qubits, with error rates improving steadily. The Jiuzhang photonic computer claimed quantum computational advantage in 2020—for a problem that was, admittedly, not particularly useful, but technically impressive regardless.[26]

Where America throws seeds broadly and hopes for a forest, China plants specific crops on schedule and harvests according to plan. The advantage is obvious: resource mobilisation at a scale and speed that no democratic funding cycle can match. The risk is equally obvious: centralised direction produces the planned breakthroughs and misses the unplanned ones. The internet was not on anyone's five-year plan. Neither was penicillin. The question of whether you can schedule a scientific revolution remains genuinely open.

The private sector has kept pace. QuantumCTek—the commercial arm of Pan Jianwei's quantum communication research group—listed on the Shanghai STAR Market in 2020, becoming the world's first publicly traded quantum communications company.[27] Its systems are deployed in banking networks and government communications across China. Origin Quantum, founded in 2017, operates China's dominant quantum cloud platform.[28] Not in a laboratory. In banks.

Europe: Collaborative Complexity

Europe has committed €12 billion including national programmes—which sounds substantial until you notice that China announced $10 billion for a single campus in Hefei.[2]

The EU Quantum Flagship launched in 2018 with €1 billion—at the time the largest civilian quantum investment in history.[2] Then China announced $10 billion, the United States passed the NQI Act, and the EU's billion looked like a rounding error. The Flagship has since grown to €12 billion including national programmes. Individual member states layered their own investments: the UK at £3.5 billion, Germany at €3.2 billion, France at €2.4 billion, with the Netherlands leading quantum internet research through QuTech at Delft—the world's most credible quantum internet programme.[29]

The commercial ecosystem has genuinely matured. Oxford Quantum Circuits delivers superconducting systems commercially. IQM has shipped quantum computers to Germany and Spain. Pasqal's neutral atom systems are in production. Alpine Quantum Technologies ships ion trap systems.[30] These are products, not pilots. The European quantum industry exists in a way it did not five years ago.

The challenge—and the charm, and the occasional nightmare—is that building a quantum superpower by committee requires all twenty-seven members to agree on direction, timeline, standards, and who gets the credit. Europe has a genuine theoretical physics tradition, a collaborative culture that produces high-quality science, and a useful obsession with standards and interoperability. What it does not have is an equivalent to Sand Hill Road. European quantum startups raising Series B rounds routinely discover that the serious cheques are written in Palo Alto or Beijing, not Munich or Amsterdam.

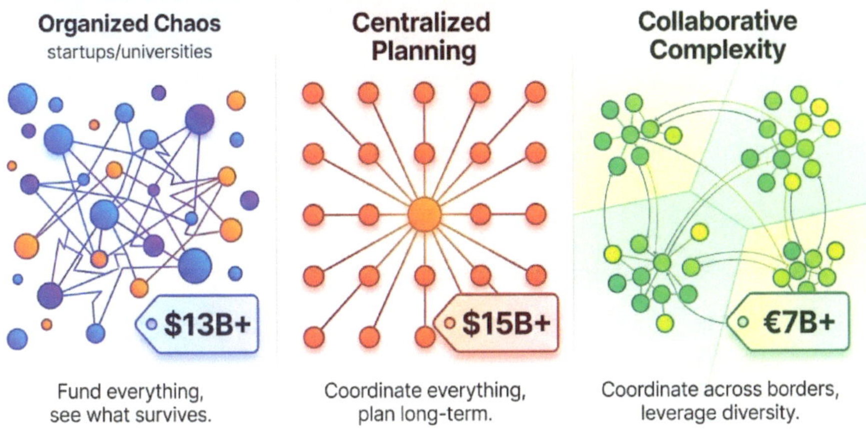

Three strategies. One race. We'll know which worked in 2035.

Europe is trying to win a Formula 1 race by committee, with every committee member holding veto power over tyre choice. But it is worth noting what the committee has actually produced: real deployments, real standards, and EuroQCI—a quantum communication network now in active deployment connecting member states.[31] The committee is getting things built. Just more slowly than the members would like, and at greater cost than the budget originally suggested—which is how committees usually work.

India's Bet: Speed Over Perfection

In 2023, India approved the National Quantum Mission with a budget of approximately ₹6,000 crore—roughly $730 million—to be deployed over eight years. By the standards of what China is spending on a single campus in Hefei, this is modest. By the standards of what most countries spend on emerging technology, it is serious.

The NQM targets four major goals: indigenous quantum computers, quantum communication, quantum sensing, and a talent pipeline. C-DAC, the IITs, and IISc form the research backbone. The strategic logic is different from both the US and European approaches: India is not trying to lead the hardware race. It's trying to build the expertise base that ensures Indian institutions—and Indian engineers—are not permanently dependent on foreign quantum infrastructure.

Whether the timeline is achievable is debated by researchers inside the programme. What isn't debated: the alternative—having no quantum programme at all—is worse. A nation of 1.4 billion people and a rapidly growing technology sector cannot afford to have zero sovereign quantum capability when the technology matures.

India: The Sleeping Giant Wakes Up

There's a question that doesn't get asked enough in Western coverage of the quantum race: what happens to the 1.4 billion people who live in neither China, America, nor Europe?

India answered that question in April 2023, when the Union Cabinet approved the National Quantum Mission with a budget of ₹6,003 crore—approximately $730 million—to be deployed over eight years through 2031. [43] The announcement didn't make international headlines. Quantum announcements from India rarely do. That is a mistake.

Here's why India's bet is different from everyone else's:

America is trying to win the hardware race. China is trying to win everything simultaneously. Europe is trying to build standards and collaborative infrastructure. India, with characteristic pragmatism, is trying to do something more specific: ensure that Indian engineers, Indian institutions, and Indian companies are not permanently dependent on foreign quantum infrastructure when the technology matures.

The NQM has four target areas: indigenous quantum computers in the 50–1,000 qubit range by 2031, satellite-based quantum communication over 2,000 kilometres, quantum sensing and metrology systems for navigation and medical imaging, and—critically—the workforce. The mission explicitly funds quantum education and research fellowships, acknowledging what most quantum programmes quietly ignore: hardware without trained humans is expensive furniture.

The institutional backbone involves C-DAC, the IITs, IISc, TIFR, and the newly created Quantum Computing Applications Lab. IIT Madras,

IIT Bombay, and IISc Bangalore already have active quantum research groups whose output competes internationally. India produces approximately 200 quantum-related PhDs annually[44]—a small number by Chinese or American standards, but a foundation that the NQM is designed to multiply rapidly.

Is the timeline achievable? Researchers inside the programme debate this with the specific candour that Indians reserve for honest conversations about their own institutions. The bureaucratic coordination required—across multiple scientific departments, multiple institutions, multiple states—is genuinely challenging. India's history with large technology programmes has been uneven.

But the alternative—having no programme—is clearly worse. A $730 million commitment over eight years is modest by China's standards. It is serious by any other measure. And India has a strategic asset that China and America both lack: it sits between them. India's quantum programme can choose to collaborate with Western partners (as it has historically in space research) without the geopolitical complications that define US-China quantum competition. That positioning is worth something.

Pan Jianwei returned to China in 2001 and spent eight years building toward a satellite. The people who started India's NQM in 2023 are doing something similar, if quieter. The question is whether eight years of focused effort will be enough to put India in the room when the quantum age matures.

If history is any guide, underestimating India's ability to surprise is a consistent mistake.

When Quantum Became an Asset Class

In October 2021, IonQ began trading on the New York Stock Exchange, becoming the first pure-play quantum computing company to go public.[20] Its implied valuation at listing was $2 billion. Its actual revenue that year was $2.1 million. Its most powerful quantum computer had 32 qubits. Investors did not care. They were betting on what quantum computing could do in 2031.

This is not a story about irrational exuberance, though there was plenty. It is a story about what happens when institutional money decides that being late to a technology transition is more expensive than being early and wrong. More than $12 billion in venture capital has entered the sector since 2018.[12] Some of those bets are now paying off. Some are running out of road. The ones running out of road are not the interesting story.

But the more revealing story is not the startups. It is the enterprises—and what they tell us about where serious money thinks this is going.

In 2021, Honeywell spun out its quantum division and merged it with Cambridge Quantum Computing to form Quantinuum—instantly the largest integrated quantum computing company by technical breadth.[32] By 2023, Quantinuum had raised funding at a $5 billion valuation. This was not a startup making promises. This was an industrial conglomerate deciding quantum computing was real enough to build a serious company around it. When Honeywell—a company that makes jet engines and building controls— reorganises itself around quantum computing, the technology has crossed a threshold no press release can fake.

The more telling signal came from the enterprises.

JPMorgan published peer-reviewed research on quantum algorithms for derivatives pricing.[33] Goldman Sachs built a quantum computing team and ran Monte Carlo simulation experiments—the mathematical engine behind most financial risk modelling.[34] Airbus ran quantum optimisation for aircraft loading problems, where fractional fuel savings translate to hundreds of millions annually.[35] Boehringer Ingelheim signed a multi-year partnership with Google Quantum AI for molecular simulation.[36]

None of these programmes had delivered commercial products by 2026. That is not the point. The point is that JPMorgan and Goldman Sachs are not in the business of funding curiosity. When they build quantum teams, it is because their analysts believe quantum computing will transform their industry within a decade—and they cannot afford to be the institution that does not understand it when that moment arrives. In Chapter 18, we'll see how this financial quantum investment intersects with AI to create something more powerful than either technology alone.

McKinsey, Boston Consulting Group, and Accenture all launched dedicated quantum computing consulting practices.[37] When quantum computing appears as a line item on McKinsey's service menu, it means CEOs are being asked about it in quarterly reviews. The consulting layer is the FOMO indicator—the moment a technology stops being something researchers talk about and starts being something CFOs worry about.

McKinsey projects $650 billion to $1.2 trillion in quantum computing value by 2035[38]—a figure that should be read as 'this is the number McKinsey uses when CFOs ask for a number,' not as a scientific forecast. It is, however, a useful signal: the range exists because this is what gets typed into boardroom presentations.

The challenge is separating companies building real capability from those mastering quantum-flavoured press releases. The physics is difficult enough that sophisticated investors sometimes cannot tell the difference. The market will eventually correct. The question is whether correction arrives before or after genuine commercial applications, and which companies survive the gap.

The Fragmented Finish Line

Ask a quantum physicist who is winning the quantum race and they will answer a question you did not ask: winning what, exactly?

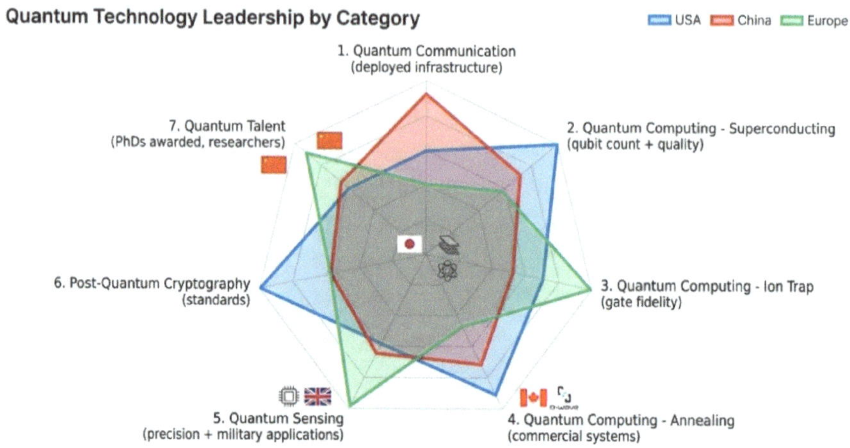

No single nation dominates all quantum technologies. Leadership is distributed across different approaches.

The answer depends entirely on which race you mean.

China has built more quantum communication infrastructure than the rest of the world combined—two operational satellites, a 4,600-kilometre terrestrial network, deployment capabilities nobody else has matched.[25]

The United States leads on quantum computing hardware; Google's error correction milestone and IBM's roadmap execution are setting the pace.[14,16]

Europe leads on quantum sensing deployment and—perhaps most importantly in the long run—the standards work that will determine how quantum networks from different countries actually talk to each other.[31]

Japan leads in quantum annealing. Australia has produced world-record coherence times in silicon qubits.[39]

Canada, working with a fraction of the budget, punches above its weight in quantum computing.

There is no single finish line. The "quantum race" is actually five separate races—computing, communication, sensing, cryptography, and talent—running simultaneously, with different leaders in each.

GLOBAL QUANTUM TECHNOLOGY LEADERSHIP: APPLICATIONS VS. REGIONS (2025)

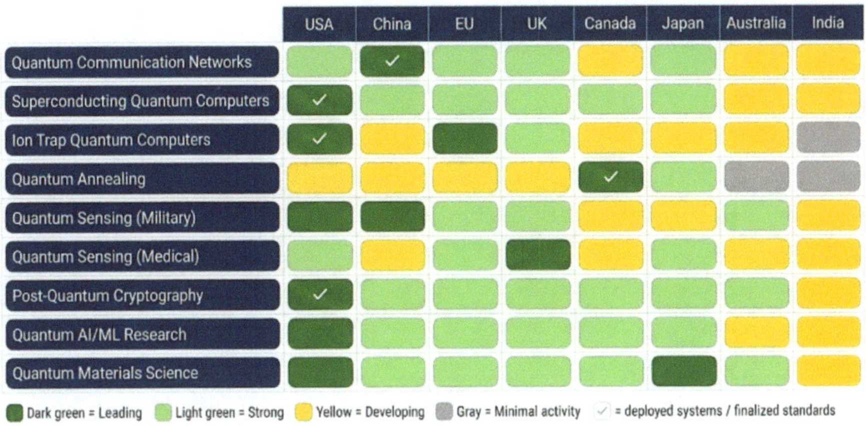

Quantum technology leadership varies by application. No single nation dominates all areas. Success requires choosing strategic priorities and building on existing strengths.

Photonic Quantum Computing

And there is a sixth race that almost nobody is talking about: the race to build quantum computers without the refrigerator.

Every quantum computer you've read about in this book operates at temperatures colder than outer space—0.015 Kelvin, achieved using dilution refrigerators the size of a wardrobe that cost between $500,000 and $2 million each. This is one of the primary reasons quantum computers will never be in your pocket. You cannot fit a dilution refrigerator in a pocket.

But there's an alternative approach that uses a completely different physical substrate: light.

Photonic quantum computers use photons—particles of light—as qubits. Photons don't need to be cold. They travel at the speed of light (which helps). They don't interact easily with their environment (which reduces decoherence). And they can be manipulated using technology that the telecommunications industry has spent fifty years perfecting—optical fibres, beam splitters, and detectors.

The trade-off: photons are hard to make interact with each other, which is precisely what you need for the entanglement that makes quantum computing powerful. Getting photons to "talk" requires extremely precise optical components and probabilistic operations that classical physics doesn't impose.

PsiQuantum, a startup that has raised over $900 million from investors including Microsoft and Goldman Sachs, is betting everything on photonic quantum computing at scale.[45] Their argument: silicon photonic chips can be manufactured using existing semiconductor fabrication infrastructure—TSMC can make photonic chips. The path to millions of qubits, they argue, runs through photonics, not superconductors.

Xanadu, a Canadian company, demonstrated Borealis in 2022—a photonic quantum processor that outperformed classical supercomputers on a specific sampling task, using 216 squeezed light modes.[46] In 2024, they published results showing continued progress in programmable photonic circuits.

Neither company has demonstrated a photonic quantum computer that solves a practically useful problem. Neither has demonstrated fault-tolerance. The approaches that have demonstrated error correction—Google's Willow, IBM's superconducting systems—all use superconducting qubits at millikelvin temperatures.

But the photonic bet is a long bet on a specific architectural advantage: if you need millions of physical qubits for fault-tolerant quantum computing (Chapter 12's 1,000:1 ratio), the only manufacturing infrastructure capable of producing millions of identical, high-quality components is the semiconductor industry. And the semiconductor industry can make photonic chips.

Whether photonic quantum computing beats superconducting is one of the genuinely open questions in the field. The answer will arrive sometime in the 2030s. The winner will look obvious in retrospect. Right now, anyone who tells you they know which approach will dominate is more confident than the evidence justifies.

Talent Gap

The talent gap is where the deficit is most unambiguous. Global PhD production in quantum-related fields has reached approximately 3,200 annually.[40] Industry demand for quantum engineers, quantum chemists, and quantum algorithm developers is estimated at 15,000–20,000 positions by 2028.[40] That gap does not close by itself. It closes through universities redesigning curricula, companies funding graduate programmes, governments making quantum education a national priority with the same urgency they applied to STEM after Sputnik. Most governments have funded the hardware. They have not yet funded the humans who will run it.

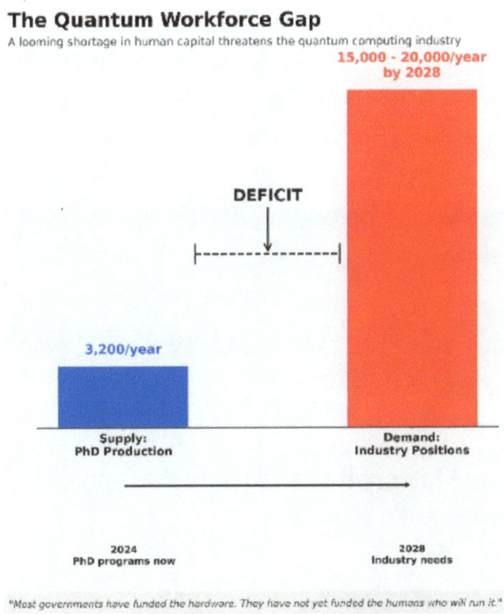

Then there is the cryptography problem, which is less a race and more a ticking clock. NIST released its post-quantum cryptography standards in 2024.[13] The current best estimate for when a quantum computer could break RSA-2048 encryption runs to the early 2030s.[41] Roughly 60% of critical systems are estimated to still be running on vulnerable encryption. Migrating those systems is slow, expensive, and unglamorous work that generates no share price movement. It is also, depending on how the hardware timeline plays out, the most important IT security project in history.

The nations that have not begun serious migration programmes are, in intelligence community language, "harvest now, decrypt later" targets—adversaries may already be collecting encrypted communications today, storing them, and waiting for the quantum decryption capability that will unlock them retroactively. Every sensitive communication sent over vulnerable encryption in the next five years may eventually be readable. That window is not closing fast enough.

The Thing Nobody Is Talking About

In 1945, the atomic bomb ended the Second World War and began the Cold War simultaneously. The technology arrived decades before the governance structures capable of managing it. The Nuclear Non-Proliferation Treaty was not signed until 1968—twenty-three years after the first bomb was used in combat. The International Atomic Energy Agency was established in 1957: twelve years after Hiroshima.

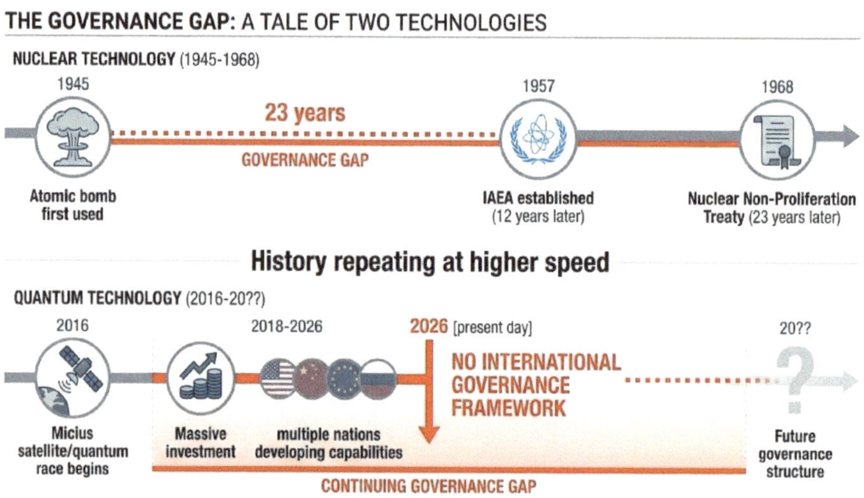

> *We built the thing first. We figured out how to govern it later. The costs of that sequencing are visible in every nuclear-armed standoff of the past eighty years.*

Quantum technology is following the same pattern, at higher speed.

Dozens of nations are developing quantum capabilities simultaneously, and nobody has agreed on what the rules are. There are no universal standards for quantum communication protocols. No export control framework with genuine buy-in from the major players. No international body that can verify quantum capabilities the way the IAEA verifies nuclear ones. Not even an agreed definition of what would constitute a 'quantum weapons capability.' The technology is outrunning the vocabulary, let alone the governance.

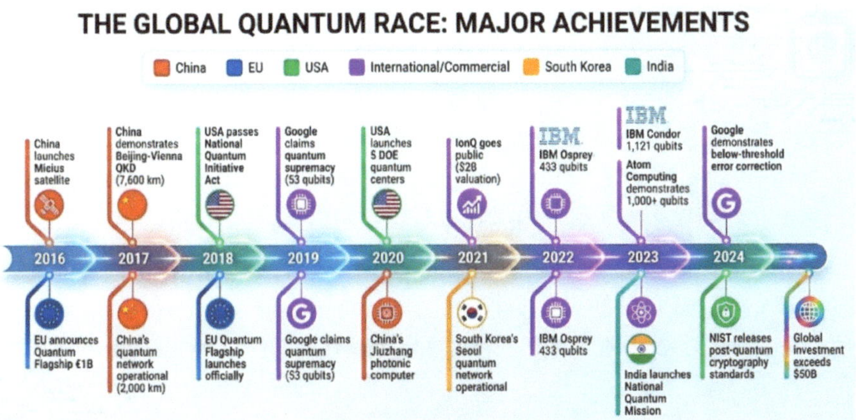

Export controls on quantum hardware have tightened significantly. Quantum computers above certain qubit thresholds now require export licences. Cloud access to advanced quantum systems is geographically restricted. Certain quantum sensing technologies are classified as dual-use and require government approval to transfer.[42] Some nations have responded with reciprocal controls on cryogenic equipment, precision lasers, and specialised components. The quantum scientific community still collaborates extensively—papers are published, conferences are international, researchers move between institutions. But the walls are higher than they were five years ago, and rising.

Here is the uncomfortable truth: every governance failure that preceded a nuclear near-miss, every arms race that didn't have to happen, every deterrence crisis that caused decades of anxiety—all of it began with technology arriving faster than the vocabulary for managing it. We knew what the bomb could do. We just hadn't decided what we wanted to do about that. Quantum computing is not a bomb. But the pattern of arriving before the framework is exactly the same. The difference is we have a warning this time. History has sent us a memo.

What nobody has built is an international governance structure adequate to the technology being created. The decisions being made right now—in 2026—in government funding agencies, corporate R&D centres, and university hiring committees—will determine technological leadership through the 2040s. But they are being made without agreed rules, without shared standards, and without any mechanism for the kind of transparency that might prevent a quantum capability demonstration from being interpreted as an act of aggression.

Pan Jianwei left Vienna in 2001, built a satellite that changed the strategic calculus of every major nation on Earth, and in doing so started a race that nobody had a framework to manage. The quantum race has a hundred well-funded participants and no referee.

That is the thing nobody is talking about. And the window to talk about it—before the capabilities outrun the governance structures entirely—is shorter than the $60 billion investment suggests anyone believes.

Turn the page.

References

[1] U.S. Congress (2018). National Quantum Initiative Act. Public Law 115–368, signed December 21, 2018.

[2] European Commission (2018). Quantum Technologies Flagship: A €1 billion initiative. Quantum Flagship Programme Launch Document.

[3] UK Research and Innovation (2024). National Quantum Strategy. UKRI Programme Report.

[4] Cabinet Office of Japan (2022). Quantum Innovation Initiative. Council for Science, Technology and Innovation.

[5] Department of Science and Technology, India (2023). National Quantum Mission: Implementation Guidelines. Government of India.

[6] Government of Canada (2023). National Quantum Strategy. Innovation, Science and Economic Development Canada.

[7] Ministry of Science and ICT, South Korea (2023). Quantum Technology Development and Promotion Act. MSIT.

[8] Department of Industry, Science and Resources, Australia (2023). National Quantum Strategy. Australian Government.

[9] Kania, E. B., & Costello, J. (2021). Quantum Hegemony? China's Ambitions and the Challenge to U.S. Innovation Leadership. Centre for a New American Security.

[10] National Quantum Coordination Office (2024). National Quantum Initiative Annual Report. Office of Science and Technology Policy.

[11] McKinsey Global Institute (2024). Quantum Technology Monitor. McKinsey & Company.

[12] PitchBook Data (2024). Quantum Computing: Venture Capital Activity 2018–2024. PitchBook Market Report.

[13] NIST (2024). Post-Quantum Cryptography: Federal Information Processing Standards. National Institute of Standards and Technology.

[14] IBM Quantum (2023). IBM Quantum Development Roadmap. IBM Research Technical Report.

[15] Arute, F., et al. (2019). Quantum supremacy using a programmable superconducting processor. Nature, 574, 505–510.

[16] Acharya, R., et al. (2024). Quantum error correction below the surface code threshold. Nature, 614, 676–681.

[17] Castelvecchi, D. (2022). Majorana particle claim retracted after reanalysis of key data. Nature, 603, 21–22.

[18] Aghaee, M., et al. (2025). InAs-Al hybrid devices for topological quantum computing. Nature, 630, 329–334.

[19] Amazon Web Services (2020). Amazon Braket: Quantum computing service. AWS Product Documentation.

[20] IonQ (2021). IonQ becomes first publicly traded pure-play quantum computing company. Company Press Release, October 1, 2021.

[21] Bluvstein, D., et al. (2024). A quantum processor based on coherent transport of entangled atom arrays. Nature, 626, 58–65.

[22] Rigetti Computing (2024). Quantum computing systems and cloud access. Company Technical Documentation.

[23] PsiQuantum (2024). Photonic quantum computing development programme. Company Announcements and Funding Disclosures.

[24] South China Morning Post (2017). China building world›s biggest quantum research facility. September 11, 2017.

[25] Chen, Y. A., et al. (2021). An integrated space-to-ground quantum communication network over 4,600 kilometres. Nature, 589, 214–219.

[26] Zhong, H. S., et al. (2020). Quantum computational advantage using photons. Science, 370(6523), 1460–1463.

[27] QuantumCTek (2020). Initial public offering prospectus. Shanghai STAR Market Filing.

[28] Origin Quantum (2023). Quantum cloud computing platform and services. Company Technical Documentation.

[29] Pompili, M., et al. (2021). Realization of a multinode quantum network of remote solid-state qubits. Science, 372(6539), 259–264.

[30] European Quantum Industry Consortium (2024). Commercial quantum systems deployment report. EQUIC Industry Survey.

[31] European Commission (2024). European Quantum Communication Infrastructure implementation status. EU Digital Programme Report.

[32] Quantinuum (2021). Formation announcement and strategic vision. Company Press Release, June 2021.

[33] Herman, D., et al. (2023). Quantum computing for finance. Nature Reviews Physics, 5, 450–465.

[34] Egger, D. J., et al. (2020). Quantum computing for Finance: state of the art and future prospects. IEEE Transactions on Quantum Engineering, 1, 3101724.

[35] Stollenwerk, T., et al. (2020). Quantum annealing applied to de-conflicting optimal trajectories for air traffic management. IEEE Transactions on Intelligent Transportation Systems, 21(1), 285–297.

[36] Boehringer Ingelheim (2021). Quantum computing partnership for pharmaceutical research. Company Press Release.

[37] McKinsey & Company (2023). Quantum computing consulting practice launch. McKinsey Digital Practice Announcement.

[38] McKinsey & Company (2024). Quantum computing value creation: $650 billion to $1.2 trillion by 2035. McKinsey Global Institute Report.

[39] Yang, C. H., et al. (2020). Silicon qubit fidelities approaching incoherent noise limits via pulse engineering. Nature Electronics, 3, 469–474.

[40] Quantum Economic Development Consortium (2024). Building the quantum workforce: demand projections 2024–2028. QEDC Workforce Report.

[41] Gidney, C., & Ekerå, M. (2021). How to factor 2048-bit RSA integers in 8 hours using 20 million noisy qubits. Quantum, 5, 433.

[42] U.S. Department of Commerce (2023). Export Administration Regulations: Quantum Computing Controls. Bureau of Industry and Security Federal Register.

[43] Department of Science and Technology, India (2023). National Quantum Mission: Approved by Union Cabinet, April 19, 2023. Press Information Bureau, Government of India.

[44] Quantum Economic Development Consortium (2024). Global quantum workforce production by nation, 2024. QEDC Workforce Analysis Report.

[45] PsiQuantum (2024). Photonic quantum computing development programme. Company announcements and SEC filings, 2022–2024.

[46] Madsen, L. S., et al. (2022). Quantum computational advantage with a programmable photonic processor. Nature, 606, 75–81.

PART V

THE QUANTUM RECKONING

Every quantum computer requires a classical computer to function. The technology does not arrive alone. It arrives with whoever built it, whoever funded it, and whatever they decided it was for. Choosing what to build next is not a technical decision.

CHAPTER

18

Quantum Machine Learning

When the two most powerful technologies of the century started helping each other

The Discovery Nobody Expected

In October 2024, researchers at Google DeepMind published something that made quantum physicists uncomfortable. An AI system had discovered a new quantum error correction code—a mathematical structure for protecting quantum information from noise—that was more efficient than anything humans had designed in thirty years of trying.[1] The AI didn't understand why the code worked. It couldn't explain its reasoning. It just searched through billions of possible configurations and found something better than the best human experts had achieved. Which means, for the first time in the history of physics, scientists are using a tool they cannot interrogate. They have the answer. They do not have the understanding. Two months later, the same research group announced the AI had discovered three more quantum codes, each optimised for different types of quantum hardware.[2]

The reason this matters is specific. Quantum computing's central unsolved problem is errors—quantum information is so fragile that noise destroys it faster than most computations can complete. The mathematical codes that protect quantum information from noise are extraordinarily difficult to design; the best human researchers have been working on them since the 1990s and have found perhaps a dozen genuinely good solutions.

> *The AI found a better one in weeks. And then found three more.*

The AI solved it in weeks.

Let me make sure you feel the weight of that sentence. Quantum error correction is one of the hardest problems in theoretical physics. The world's best human researchers—people who spent their careers thinking about nothing else—had thirty years and a dozen genuinely good solutions to show for it.

An AI system, given no special insight into quantum physics, no intuition about which approaches were promising, not even the ability to explain its own reasoning, searched through billions of possible configurations and found something better. Then found three more.

Science has never had a tool like this before.

> *We have the answer. We don't have the understanding. The AI is the oracle, and we're the ancient Greeks—receiving pronouncements we can follow but cannot interrogate.*

It also should feel enormously exciting. Because it means the pace of quantum discovery is about to accelerate in ways that depend less on how fast human researchers can think and more on how much computing power we can direct at the problem.

Here is the loop this creates:

Better error correction enables fault-tolerant quantum computers. Fault-tolerant quantum computers enable better quantum AI. Better quantum AI discovers better error correction. The wheel turns, and each rotation makes the next one faster.

> *We've created a feedback loop between the two most powerful computational technologies ever developed.*

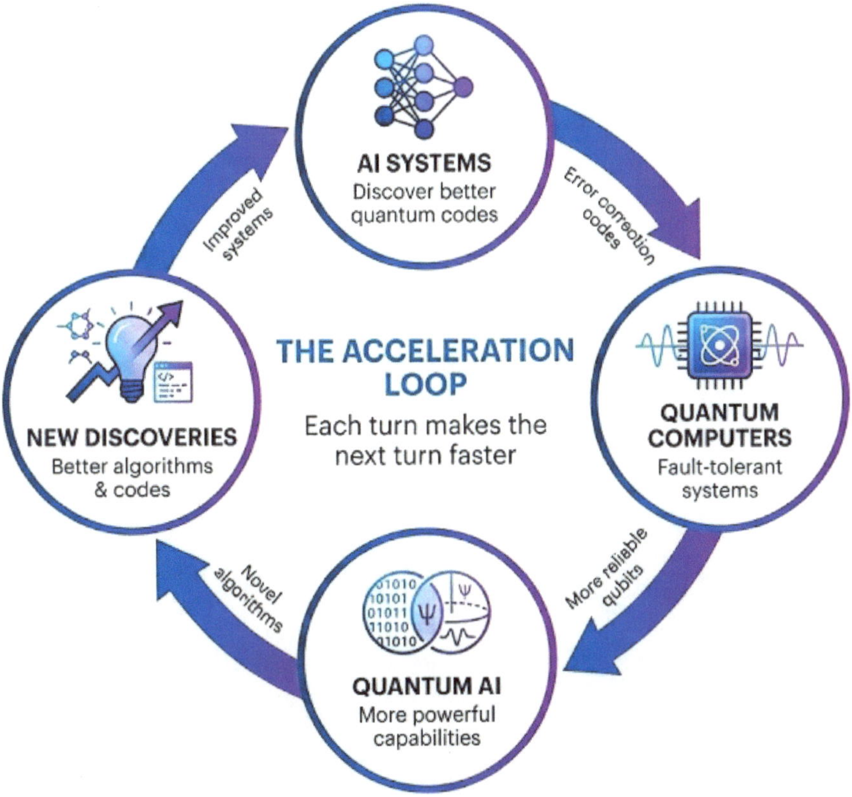

And nobody's quite sure where it leads.

The Human Who Hit the Wall

To understand why the loop matters, you need to know about six months of wasted work.

In 2019, a PhD student at a leading university spent six months training a quantum neural network. The architecture was sound. The mathematics were correct. The circuit compiled without errors. She ran thousands of training iterations—each one adjusting the parameters of the quantum circuit, searching for configurations that made better predictions.

The network learned absolutely nothing.

Not slowly. Not partially. Nothing. After six months and thousands of iterations, its predictions were no better than random guessing. She ran diagnostics. Everything appeared to be working. The gradients—the signals that tell a neural network which way to improve—were there in theory. But in practice they were so small, so swamped by quantum noise, that the training process could not detect them.

She had stumbled into what researchers now call the barren plateau problem.[3]

NEURAL NETWORK TRAINING LANDSCAPES

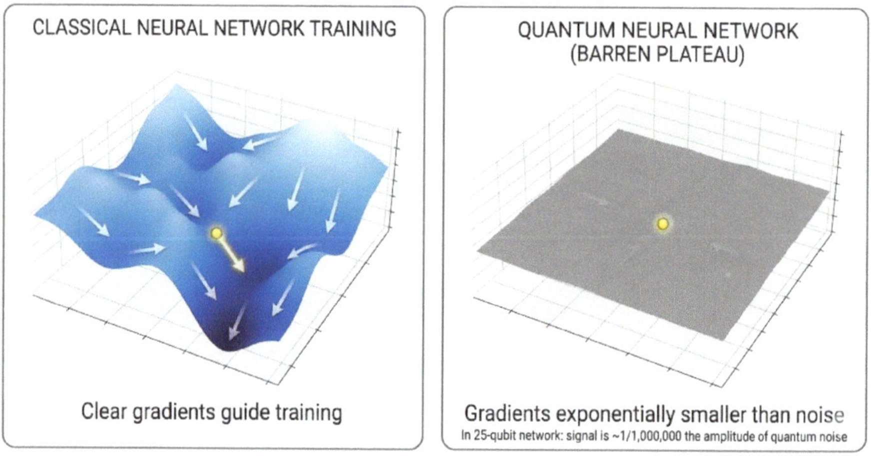

Imagine training a neural network as navigating a mountain landscape, looking for the lowest valley. Classical neural networks work because the landscape has features—slopes, ridges, gradients you can follow downhill. The barren plateau problem is what happens when you scale up a quantum neural network: the landscape flattens.

The mathematics behind this are brutal. In 2018, a research team at Google proved that as you add qubits to a quantum neural network, training signals get exponentially weaker.[3] In a 25-qubit quantum neural network, the gradient signals you need to detect are roughly one-millionth the amplitude of quantum noise. You're searching for a whisper inside a hurricane.

The MIT PhD student's quantum network had exactly 25 qubits. The gradients were mathematically present but physically undetectable. Six months of training, zero learning. She switched to a classical network and graduated.

That last sentence sounds like defeat. It is also the setup for everything that follows. The barren plateau problem meant that humans, building quantum AI by hand using classical intuition, had hit a ceiling. The networks that should have been powerful became trainable only when kept tiny. Every approach that ought to have scaled refused to.

Researchers published workarounds—keep the circuits shallow, choose certain architectures that preserve gradients better, use hybrid designs where quantum handles specific bottlenecks. They worked around the problem. They did not solve it.

The AI that discovered better quantum error correction codes in 2024 was not working around the barren plateau problem. It was not relying on intuition, or gradient signals, or any of the training mechanisms that hit ceilings. It searched. It operates by a different mechanism entirely—not intuition, not analogy, not the trained pattern-recognition that human researchers rely on. It searches spaces humans find impenetrable. And in those spaces, it is finding things.

This is the ground floor of the loop. The reason AI can help quantum computing in ways humans cannot is that the two failure modes do not overlap. Humans hit cognitive ceilings. AI hits resource ceilings. When you combine them, the ceiling moves.

The Acceleration Loop

To understand the feedback loop, it helps to see it as two turns of a wheel—each one making the other possible.

The best way to understand the quantum-AI feedback loop is to think about two people stuck in quicksand.

Person A can't move their legs (they're trapped by limited qubits and decoherence). Person B can't reach the rope they need to pull themselves out (they're limited by classical computing power for certain problems). But if Person A can throw something to help Person B grab higher, and Person B can then pull Person A partially free, and Person A can now reach something to help Person B further... you have a feedback loop where each small gain makes the next gain possible.

That's the quantum-AI loop. Neither technology needs to be fully mature to start helping the other. Each small improvement cascades.

Turn one: AI improves quantum computers. Turn two: Better quantum computers improve AI. The wheel spins. Each rotation makes the next one faster.

Here's what that looks like in practice.

The first turn is AI helping quantum. Quantum computers are extraordinarily difficult to operate. Every qubit drifts—temperature fluctuations, electromagnetic noise, and even cosmic rays continuously push qubits away from the states required for computation. Calibrating a quantum computer manually is roughly equivalent to tuning a concert grand piano by ear, continuously, while someone keeps changing the temperature of the concert hall.

By 2026, every major quantum computer uses machine learning for real-time calibration—reinforcement learning finding optimal control pulses, neural networks compensating for crosstalk between qubits, predictive models identifying when a qubit is about to fail.[4] The gap between human-tuned and ML-tuned quantum systems is now large enough that manual calibration is no longer seriously attempted on production hardware.

Beyond calibration, AI now routinely optimises quantum circuits—the sequences of operations that execute quantum algorithms. Circuits designed by humans and then optimised by AI systems at Google and IBM achieve 20–30% fewer gate operations than the original human design.[5] Fewer gates means fewer opportunities for errors. Fewer errors means more reliable computation. This is AI making quantum computers better without changing a single piece of hardware.

Then there is the deeper contribution: AI finding quantum algorithms that humans never imagined. In thirty years of quantum computing research, human researchers have discovered perhaps a dozen genuinely useful quantum algorithms—Shor's factoring algorithm, Grover's search, VQE for chemistry, QAOA for optimization, a handful of others.[6] The search is hard because the space is vast—the number of possible quantum circuits grows exponentially with the number of qubits, far beyond what human intuition can navigate.

AI does not navigate by intuition. It searches. And it is finding things.

We opened this chapter with one example. By 2024, the examples were accumulating. Researchers at Quantinuum demonstrated an AI discovering quantum circuit optimizations that their own engineers could not explain until they reverse-engineered the solution after the fact.[7] Machines finding answers. Humans working out why they are answers afterward.

This is new in the history of physics.

The second turn of the wheel runs in the opposite direction: quantum helping AI.

This is subtler and currently smaller in practical impact—but potentially the more transformative of the two. Machine learning works by mapping data into high-dimensional mathematical spaces where patterns become detectable. A classical computer creates these spaces through mathematical transformations. A quantum computer can create spaces that are exponentially larger, using quantum superposition to represent many states simultaneously.[8]

This is quantum kernel machine learning—using a quantum processor to map data into a quantum feature space, then using classical algorithms to find the patterns that live there. Some patterns that are invisible in classical feature spaces appear clearly in quantum ones. In molecular data—the kind that determines whether a drug candidate will bind to a protein or whether a new material will conduct electricity—quantum feature spaces can reveal structure that classical methods systematically miss.

By 2026, researchers have demonstrated 5–15% accuracy improvements in specific molecular prediction tasks using quantum kernel methods.[9] That range sounds modest. In drug discovery, a 10% improvement in predicting whether a molecule will bind to a target protein means eliminating hundreds of failed experiments before they begin. At the scale pharmaceutical companies operate—testing thousands of candidates to find one drug—those percentage points represent years of time and hundreds of millions of dollars.

The improvements are not revolutionary in the laboratory. They are revolutionary in the economics.

That is the loop. And it is already moving—AI making quantum computers better. Quantum computers making AI more capable. Each turn of the wheel makes the next turn faster.

What Actually Works (And What Doesn't)

By 2026, the quantum-AI loop has moved from laboratory demonstration to commercial pilot programmes. Here's what's working, what isn't, and what the numbers actually show.

QUANTUM-AI APPLICATIONS: REALITY CHECK

Application	Classical Approach	Quantum-AI Hybrid	Improvement	Status (2026)
Molecular Property Prediction	Neural networks on historical datasets	Quantum feature spaces + classical ML	**5–15%** accuracy gain	Commercial deployments
Drug Binding Affinity	Molecular docking + ML scoring	Quantum simulation + AI pattern recognition	**7–12%** better predictions	Pharma pilots
Portfolio Optimization	Classical optimization algorithms	QAOA + classical refinement	**3–8%** better returns	Limited commercial use
Quantum Circuit Design	Human-designed circuits	AI-optimised circuit synthesis	**20–30%** fewer gates	Standard practice
Materials Discovery	DFT + machine learning	Quantum simulation + AI	**2–3×** faster discovery cycles	Transition to commercial
Fraud Detection	Classical anomaly detection	Quantum kernels + ML	**5–10%** new pattern detection	Pilot testing

Commercial deployments here refers to pilot-stage revenue-generating programmes; general commercial availability varies by provider.

Key: Comm. = Commercial, Deploy = Deployment, Transit = Transitioning

When I write "5–15% accuracy gain," I mean: in controlled comparisons on specific tasks, the quantum-AI hybrid system produced results that were 5–15% more accurate than the best classical-only approach. Not "quantum is 15% better at everything." Not "quantum is 10 times faster." A 5–15% improvement in predicting whether a drug candidate will bind to its target protein, applied at the scale pharmaceutical companies operate—thousands of candidates screened before synthesis—is worth hundreds of millions of dollars in avoided failures.

Small percentages. Large consequences. That's the pattern.

Real pharmaceutical results from IBM's pilot programme:[10]
Blood-brain barrier prediction: 81% → 87% (+6%)
Binding affinity estimation: 73% → 82% (+9%)
Toxicity screening: 85% → 91% (+6%)

And as of 2026, multiple applications have paying customers. This is no longer theoretical.

Those 6–9% improvements eliminate hundreds of failed drug candidates early in the pipeline. At pharmaceutical industry scale, this translates to eliminating millions of dollars in wasted synthesis and testing costs per programme.

The Workarounds That Actually Work

The barren plateau problem is not solved. But the loop is already working around it.

By 2026, researchers have developed three strategies that make quantum neural networks trainable in practice:

Strategy 1: Stay Shallow

Use only 3–5 quantum layers instead of deep networks. Barren plateaus get exponentially worse with depth. A 5-layer quantum circuit remains trainable. A 25-layer circuit becomes impossible.

Strategy 2: Smart Circuit Design

Certain architectures preserve gradients better than others. Nobody fully understands why—these were discovered empirically through trial and error, not derived from theory. The effective architectures share a pattern: local connectivity between qubits, symmetric gate structures, careful initialization of parameters.

Strategy 3: Hybrid Architecture

This is the approach that has worked in commercial deployments. Use quantum layers only where they provide specific advantages. Route everything else to classical systems.

A pharmaceutical company uses this approach in production: A classical neural network processes most molecular data—extracting chemical features, comparing structural patterns, applying learned weights from millions of training examples. When the pipeline encounters quantum chemistry data—electron configurations, molecular orbital calculations, the energy landscapes that determine whether a molecule actually binds to a protein—that data routes to a single 8-qubit quantum layer. The quantum circuit computes features that classical networks find expensive to represent. Results flow back to the classical network for final prediction.

The outcome: 11% accuracy improvement over classical-only approaches, achieved with an 8-qubit quantum layer doing one specific job.[10]

The quantum layer isn't doing most of the work. It's handling the one specific task where quantum feature spaces help. That's the pattern everywhere in 2026: quantum neural networks aren't replacing classical AI. They're specialized processing units for specific bottlenecks.

What works now:

- Small quantum networks (4–12 qubits, 3–5 layers) for quantum chemistry data—Hybrid architectures with quantum handling molecular features
- Targeted deployment where quantum advantage is measurable

What doesn't work yet:

- Deep quantum networks (barren plateaus remain unsolved)
- General image/text classification (classical networks perform better)—Large-scale problems requiring 50+ qubits The quantum neural network revolution isn't canceled. It's narrower, more specialized, and more difficult than the hype suggested. But it's working where it counts.

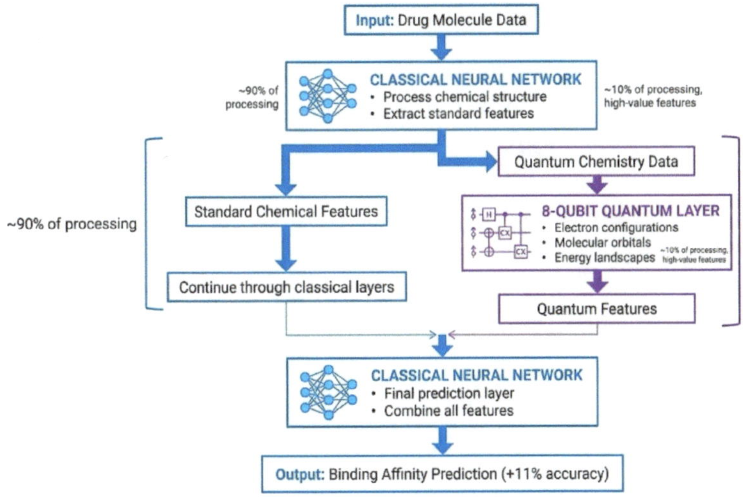

When Goldman Sachs and Boehringer Joined the Loop

You can track the maturity of a technology by watching which institutions start spending money on it before it is ready.

In the early 1990s, pharmaceutical companies started building genomics teams years before genomic medicine was commercially viable. The companies that did that work early had decisive advantages when the technology arrived. The ones that waited spent the next decade trying to catch up.

Quantum-AI is at the genomics-in-1993 moment. And the institutions that recognise it are already in the room.

For Goldman Sachs, the application is Monte Carlo simulation—the mathematical engine behind most financial risk modelling. Monte Carlo methods work by running thousands of random simulations of possible futures and finding the statistical distribution of outcomes. Quantum algorithms can, in principle, speed up Monte Carlo by quadratically—meaning the same accuracy in square-root of the time. A quantum-AI hybrid that achieves even 30% of the theoretical speedup on real hardware still saves computation time that translates to real money at Goldman's scale.

We mentioned in Chapter 17 that JPMorgan and Goldman aren't funding curiosity experiments. Here's what they're actually building. JPMorgan published peer-reviewed research on quantum algorithms for derivatives pricing—running experiments on IBM's quantum network and documenting results in academic papers.[11] Monte Carlo simulation—the mathematical engine behind most financial risk modelling. Monte Carlo methods work by running thousands of random simulations of possible futures and finding the statistical distribution of outcomes.[12] Quantum algorithms can, in principle, speed up Monte Carlo by quadratically—meaning the same accuracy in square-root of the time. A quantum-AI hybrid that achieves even 30% of the theoretical speedup on real hardware still saves computation time that translates to real money at Goldman's scale.

Airbus ran quantum optimization programmes for aircraft loading and routing—problems where finding optimal solutions across millions of variables directly affects fuel consumption and operational efficiency.[13] Boehringer Ingelheim, one of the world's largest pharmaceutical companies, signed a multi-year partnership with Google Quantum AI specifically to accelerate molecular simulation for drug discovery.[14] The announcement in January 2021 was the first time a major pharmaceutical company had committed publicly to a long-term quantum computing research programme. Others followed within eighteen months.

None of these programmes have delivered commercial products yet.

This is the correct observation and entirely the wrong conclusion.

JPMorgan and Goldman Sachs are not in the business of funding science experiments. When they build quantum teams, it is because their most rigorous analysts believe quantum computing will transform their industry within a decade—and they cannot afford to be the institution that does not understand it when that moment arrives.

The cost of being early is a few teams and a few years of research budgets. The cost of being late is watching competitors price risk more accurately, discover drugs faster, and design materials you cannot replicate.

McKinsey, Boston Consulting Group, and Accenture all launched dedicated quantum computing consulting practices within the same two-year window.[15,16,17] Consulting firms are the canaries of corporate strategy—they do not build practices around academic curiosities. They build practices around what CEOs are being asked about in board meetings.

When quantum-AI appears as a McKinsey service line, the technology has officially become a CFO's problem. That transition happened in 2022. By 2026, it is a standard item in technology strategy reviews at every major financial institution and pharmaceutical company.

What the Loop Means for You

You almost certainly do not work in quantum computing. You may not work in AI research. Here is why the loop matters to you anyway.

The first drug candidates identified using quantum-AI hybrid workflows entered early-stage evaluation in 2024–2025[18], with full clinical trial data expected in the late 2020s. The candidate was identified using a hybrid workflow: classical AI screening a library of molecular candidates, quantum simulation validating the predicted binding properties of the most promising ones. The quantum component did not discover the drug. It validated predictions that classical simulation would have taken significantly longer to confirm.

The pipeline was faster—not dramatically, not yet, but consistently. And consistency is what gets investors to fund the next pipeline, and the next. If quantum-AI drug pipelines achieve the improvements that early pilot programmes suggest, the downstream effect is drugs arriving years sooner than conventional development would deliver. Over a decade, across dozens of programmes at major pharmaceutical companies, the compounding effect is significant.

For materials—the batteries in electric vehicles, the solar panels on rooftops, the structural components in aircraft—quantum simulation is the tool that makes genuinely new materials discoverable. Classical computers simulate molecular behavior by approximating quantum equations. Quantum computers solve those equations directly.

The difference matters for the specific class of materials where quantum effects determine properties: high-temperature superconductors, novel catalysts for industrial chemistry, next-generation battery chemistries. The first commercially significant quantum-discovered material has not yet been announced. When it is, it will not look like a quantum computing story. It will look like a battery story, or a clean energy story, or a manufacturing story. The quantum contribution will be invisible to most readers.

For finance, the early results from quantum-AI optimization are modest in percentage terms and significant in absolute ones. A 5–8% improvement in risk-adjusted portfolio returns, applied to billions of assets under management, is hundreds of millions of dollars annually.[11,12] The institutions capturing that improvement first will have an advantage that is difficult to close, because the advantage compounds. Better risk models attract capital. More capital funds better research. Better research produces better models.

And for your digital security—as covered in Chapter 14—the loop accelerates the timeline for quantum computers that can break current encryption standards. Better quantum error correction, delivered by AI, is one of the inputs that determines when RSA-2048 becomes vulnerable. The loop makes that clock tick faster.

The Constraint Nobody Is Fixing

There is a single structural constraint on the loop that no amount of hardware investment solves: the people who can work at the intersection of quantum mechanics and machine learning and domain applications barely exist.

The problem is structural. Quantum physics takes three to four years of specialist training to do seriously. Machine learning takes two to three. Domain science—chemistry, materials, finance—takes three to four more on top. These tracks do not currently combine in any university's degree programme, and the people who have assembled all three components informally are few enough to count. The intersection—researchers who can design quantum algorithms, interpret AI outputs, and understand the domain problem well enough to know what questions to ask—requires expertise that current academic structures are not built to produce.

A quantum physicist needs three to four years of specialist training. A machine learning researcher needs two to three. A domain scientist needs three to four more. These training tracks do not currently combine.

Industry surveys suggest that global need for quantum-AI specialists—people who can actually design and deploy hybrid quantum-classical AI systems—currently exceeds several thousand positions.[19] The pipeline of people graduating with the combined skill stack is a fraction of that. The gap is growing faster than curricula can respond.

The gap has a corollary. Someone building both skill sets right now—quantum computing and machine learning, simultaneously—is positioning themselves for roles that will be genuinely scarce and genuinely needed. The demand is coming from pharmaceutical companies, financial institutions, advanced materials manufacturers, and every government that read Chapter 17 and took it seriously. The supply is currently measured in hundreds. The demand will be measured in thousands

It is a small intersection with enormous demand approaching it from every direction.

Where the Loop Goes

In 1995, DARPA funded a small project at Stanford called the Digital Library Initiative. The goal was to organise academic literature and make it searchable. Two graduate students—Sergey Brin and Larry Page—used that project as the foundation for a search algorithm they called PageRank. Three years later, they incorporated Google.

Nobody at DARPA funded that project intending to create a company that would eventually be worth $2 trillion. Nobody predicted Gmail, Google Maps, YouTube, Android, or the transformation of advertising. The applications followed the infrastructure. The infrastructure came from basic research that seemed only modestly useful at the time.

The quantum-AI loop is the Digital Library Initiative of 2026. Not because we know what it will produce—we don't, any more than DARPA knew in 1995 what a searchable digital library would eventually become. But because the infrastructure is being laid now, by the people who are curious enough and patient enough to work at the frontier before the frontier is comfortable.

The people who will make the most transformative discoveries in quantum-AI are, right now, probably in their mid-twenties, working on problems that don't have commercial applications yet, in programmes that sound obscure, at institutions that haven't received the press coverage they deserve.

Two consequences are already visible.

The first is quantum generative chemistry: AI systems that propose molecular designs, quantum computers that simulate whether those molecules have the predicted properties, AI systems that learn from the quantum simulation results and generate better proposals. The cycle—design, simulate, learn, redesign—runs faster than any experimental workflow possible in a laboratory. The molecules being explored in these loops include candidates for drugs, materials, and industrial catalysts that would take human chemists years to consider, let alone synthesize and test.[20]

The second is the recursive improvement of quantum hardware itself. AI systems are not only discovering better error correction codes—they are now being used to design better qubit architectures, better readout protocols, better noise suppression strategies. The researchers building the next generation of quantum processors are using AI tools trained on the performance data of the current generation.

The hardware is improving partly because of the AI, and the AI is improving partly because of better hardware. The loop is not only producing applications.

It is accelerating the tools that produce the applications. This is very early. The loop demonstrated in 2024 operates at the level of tens of qubits and narrow problem domains. The loop in 2030 will operate with thousands of reliable logical qubits and a much broader range of problems. The loop in 2035, if the hardware and software trajectories of the past five years continue, will operate at a scale where the phrase "quantum advantage" becomes as unremarkable as "cloud computing"—a capability assumed to be available, used when it helps, invisible when it doesn't.

QUANTUM-AI APPLICATIONS ACROSS INDUSTRIES: RESEARCH TO COMMERCIAL DEPLOYMENT (2026)

LEGEND
○ No significant activity
◑ Early demonstrations
● Active development/deployment

	RESEARCH Fundamental studies, algorithms	PILOT/TESTING Limited deployment, prototypes	COMMERCIAL Market-ready solutions, services	
HEALTHCARE	Drug discovery Protein folding	Quantum-AI drug candidate (Phase 1 trials 2025)	◑ Molecular screening services	← First quantum-AI drug in trials: 2025
MATERIALS/ CHEMISTRY	Battery materials Catalysts	Azure Quantum Elements users	◑ Materials screening platforms	← Azure Quantum Elements: 100+ enterprise users
FINANCE	Portfolio optimization Risk modeling	Quantum kernel ML for fraud detection	◑ Hybrid optimization services	← 5-15% accuracy improvements demonstrated
CLIMATE/ ENERGY	Atmospheric modeling Carbon capture	Catalyst discovery	○ (too early)	
LOGISTICS/ OPTIMIZATION	Supply chain Routing	Hybrid quantum-classical optimization	◑ Specific optimization services	

Quantum-AI convergence moved from research to commercial deployment in multiple industries by 2026. Healthcare and materials lead adoption.

The first wave of computing digitized information. The second wave learned patterns from data. What is beginning now is different from both: a third wave that simulates physical reality itself, at the level of the quantum mechanics that governs how matter actually behaves, and learns from those simulations.

If this works at scale—and the early results suggest it will— intelligence will not just describe the world. It will compute it.

None of which answers the more uncomfortable question. The loop makes quantum-AI more powerful. It says nothing about who controls it, who benefits from it, or what happens to the industries and individuals that find themselves on the other side of the capability gap.

We have built the technology. We have not built the framework for governing it.

Turn the page.

References

[1] Romera-Paredes, B., et al. (2024). Mathematical discoveries from programme search with large language models. Nature, 625, 468–475.

[2] Google DeepMind Research Team (2024). AI-discovered quantum error correction codes for diverse hardware architectures. Google Research Blog, December 2024.

[3] McClean, J. R., Boixo, S., Smelyanskiy, V. N., Babbush, R., & Neven, H. (2018). Barren plateaus in quantum neural network training landscapes. Nature Communications, 9, 4812.

[4] Klimov, P. V., et al. (2020). Fluctuations of energy-relaxation times in superconducting qubits. Physical Review Letters, 121, 090502.

[5] Khatri, S., et al. (2019). Quantum-assisted quantum compiling. Quantum, 3, 140.

[6] Montanaro, A. (2016). Quantum algorithms: an overview. npj Quantum Information, 2, 15023.

[7] Quantinuum Research Team (2023). Machine learning optimization of quantum circuits. Quantinuum Research Blog, December 2023.

[8] Havlíček, V., et al. (2019). Supervised learning with quantum-enhanced feature spaces. Nature, 567, 209–212. [9] Schuld, M., & Killoran, N. (2019). Quantum machine learning in feature Hilbert spaces. Physical Review Letters, 122, 040504.

[10] IBM Research (2024). Quantum-classical hybrid AI for pharmaceutical applications: pilot programme results. IBM Quantum Network Technical Report.

[11] Herman, D., et al. (2023). Quantum computing for finance: state of the art and future prospects. Reviews in Physics, 8, 100091.

[12] Woerner, S., & Egger, D. J. (2019). Quantum risk analysis. npj Quantum Information, 5, 15.

[13] Stollenwerk, T., et al. (2020). Quantum annealing applied to de-conflicting optimal trajectories for air traffic management. IEEE Transactions on Intelligent Transportation Systems, 21(1), 285–297.

[14] Boehringer Ingelheim & Google Quantum AI (2021). Partnership for quantum-accelerated drug discovery. Joint Press Release, January 6, 2021.

[15] McKinsey & Company (2023). Quantum computing: An emerging ecosystem and industry use cases. McKinsey Digital Report.

[16] Boston Consulting Group (2021). The next decade in quantum computing—and how to play. BCG Henderson Institute Report.

[17] Accenture (2022). Quantum computing: Preparing for the next wave of computing. Accenture Technology Vision Report.

[18] Stokes, J. M., et al. (2020). A deep learning approach to antibiotic discovery. Cell, 180(4), 688–702.

[19] National Academies of Sciences, Engineering, and Medicine (2019). Quantum Computing: Progress and Prospects. National Academies Press.

[20] Schwaller, P., et al. (2021). Molecular transformer: a model for uncertainty-calibrated chemical reaction prediction. ACS Central Science, 5(9), 1572–1583.

CHAPTER

19

Quantum's Dark Side

What Happens When the Technology Actually Works

The Harvest

For years, intelligence agencies worldwide are reported to have been systematically collecting and storing encrypted internet traffic – not to decrypt it. Just to store it.

Not trying to decrypt it. Just collecting it. Terabytes per day. Encrypted emails. Banking records. Medical files. The classified cables of governments, the trade secrets of corporations, and the private messages of people who had no idea any of this was happening.

All of it protected by RSA encryption that classical computers cannot break in reasonable time. So they stored it. Waiting.

They called it "harvest now, decrypt later."[1]

The logic is simple and genuinely frightening. Quantum computers will eventually break RSA encryption. When that happens—the estimates run from 2030 to 2035, with classified programmes possibly ahead of public estimates—everything stored today becomes readable.

Diplomatic cables from 2016 explaining negotiation strategies that remain relevant in 2030. Corporate emails discussing intellectual property worth billions. Government communications revealing intelligence sources. Personal messages sent with the assumption of privacy.

All of it potentially exposed retroactively.

This is Y2Q—the quantum cryptography transition.

> *The harvest is operational now. The decryption capability is under development. The timeline is measured in years, not decades.*

> *We have 5–10 years to migrate critical systems to post-quantum cryptography. As of 2026, roughly 60% of critical infrastructure hasn't begun migration.*[2]

When quantum computers can break RSA, everything encrypted with current standards becomes readable. The harvest problem alone would be enough to justify this chapter. But the harvest is the most familiar risk. The others are stranger.

That's just one problem. There are others. Some more immediate. Some more disturbing.

This is the chapter about what happens when quantum computing actually works. Not the applications we hope for. The consequences we haven't prepared for.

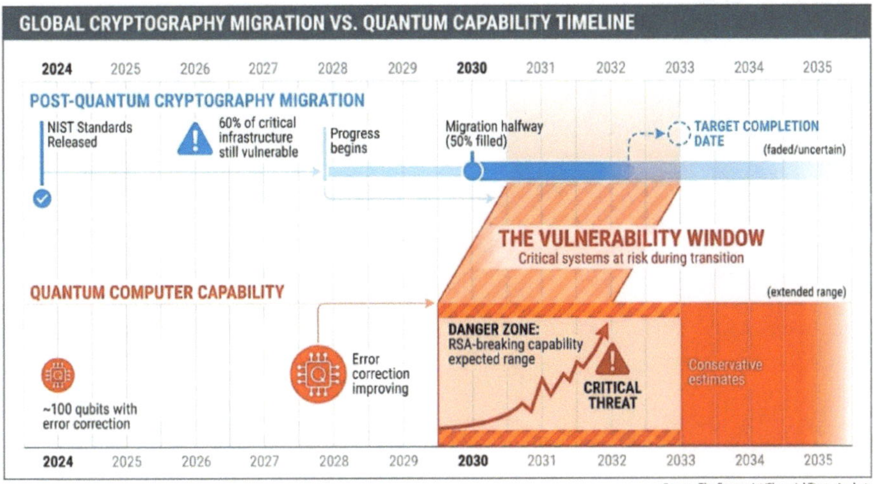

Source: The Economist/Financial Times Analysis

Y2Q: When Encryption Breaks

Encryption underpins modern civilisation in a way that only becomes visible when you try to imagine life without it. Every bank login, every medical record, every military order, every private message depends on the assumption that certain mathematical problems take longer to solve than the universe has existed. That assumption is correct—for classical computers.

> *Here is the number that keeps cryptographers awake at 3 AM. Feel free to let it keep you awake too.*

> *RSA-2048, the encryption standard protecting most of your internet traffic right now—banking, email, medical records, everything—would take a classical computer approximately 300 trillion years to break.[3] Three hundred trillion years. For context, the universe is 13.8 billion years old. The classical computer would still be running when every star in the observable universe had burned out, cooled, and gone dark.*
> *A sufficiently powerful quantum computer running Shor's algorithm could break it in hours.[4] Days at most.*

We don't have such a quantum computer yet. Conservative estimates place fault-tolerant quantum computers capable of breaking RSA-2048 around 2032–2035.[5] Optimistic projections—from a quantum computing perspective, pessimistic from a security standpoint—suggest 2030–2031. Worst case: classified quantum programmes achieve this capability earlier than public estimates suggest, don't announce it, and exploit the advantage secretly.

The clock is ticking. In the context of upgrading the encryption infrastructure of every bank, every hospital, every military system, every government database, and every piece of critical infrastructure on Earth—seven years is not a long time.

And most organizations are behaving as if they have decades to prepare.

The Migration Problem

Migrating to post-quantum cryptography is not a simple software update.

Organizations must first identify every system using vulnerable encryption—most don't have complete inventories. They must update software and hardware, including embedded systems in industrial equipment that can't be patched remotely. They must test exhaustively to ensure new cryptography doesn't break existing systems—a financial trading platform that can't complete transactions because of cryptographic incompatibility is worse than a vulnerable one. Personnel need retraining. Compatibility needs testing. And all of it needs to happen globally, across millions of systems, before a quantum computer somewhere achieves the capability. The task is enormous. The urgency has not yet matched the scale.

NIST released post-quantum cryptography standards in 2024.[6] As of 2026, implementation is proceeding slower than the threat timeline requires.

Industry surveys suggest roughly 40% of critical infrastructure remains on vulnerable encryption.[2] Migration timelines run 5–8 years at current pace. Quantum computers capable of breaking RSA: potentially operational in 4–6 years.

The math doesn't work.

What Actually Breaks

Not everything. Symmetric encryption like AES-256 remains quantum-resistant with modest key size increases.[7] But asymmetric encryption—RSA and elliptic curve cryptography, the systems that handle key exchange, digital signatures, and authentication—is vulnerable. The practical consequences are specific, and worth naming one at a time.

The practical impact:

Online banking authentication could be compromised. The secure key exchange that establishes encrypted connections becomes breakable. Banks would need to migrate every authentication system, every secure connection, every encrypted transaction—while continuing to operate normally.

Secure communications could be intercepted and decrypted retroactively. Encrypted messages sent in 2026 could be readable in 2032. For governments and corporations, this means sensitive communications from the past decade could be exposed.

Digital signatures could be forged. The cryptographic proofs that verify software updates, legal documents, and financial transactions come from legitimate sources become unreliable.

Cryptocurrency becomes vulnerable. Bitcoin and Ethereum use elliptic curve cryptography for wallet security.[8] Quantum computers could potentially break private keys, allowing theft of cryptocurrency holdings.

Military command and control systems face exposure. Encrypted orders, operational plans, intelligence communications—all potentially readable if still using vulnerable encryption when quantum computers achieve breaking capability.

Critical infrastructure control systems could be compromised. Power grids, water systems, transportation networks often use outdated encryption that's even more vulnerable than current standards.

The Harvest Is Real

The 'harvest now, decrypt later' strategy is not speculative. The NSA acknowledged it publicly in 2015. CISA has referenced it in official guidance. Security researchers who track nation-state behaviour have documented collection activity consistent with this approach.[9]

For individuals, retroactive decryption is embarrassing. For corporations, it's intellectual property theft and competitive disadvantage. For governments, it's intelligence disasters that reveal sources, methods, and strategic plans.

The Y2Q problem is solvable. Post-quantum cryptography exists and works. Migration is technically feasible. But it requires urgency most organizations haven't demonstrated. We're in a race: migrate to quantum-resistant cryptography before quantum computers become capable.

Right now, we're losing.

The Surveillance Revolution

Quantum sensors can detect things classical sensors cannot.

In 2022, researchers at the University of Birmingham deployed a quantum gravimeter in a city street. The device detected minute variations in gravitational fields caused by density differences underground. Within hours, it had found a Victorian-era tunnel that city records had lost—buried infrastructure located through gravitational anomalies classical sensors would miss.[10]

The discovery made headlines as a triumph for archaeology and urban planning. The same research group has since surveyed 12 square kilometers of Birmingham, finding 11 unmapped voids, three beneath active roadways.

The same physics that locates forgotten tunnels can map underground military facilities from aerial platforms. Detect bunkers designed to be hidden. Monitor tunnels meant to stay secret. The sensor doesn't distinguish between civilian infrastructure and military installations—it detects mass, density, gravitational anomalies.

Quantum magnetometers detect magnetic fields with sensitivity approaching fundamental quantum limits—measuring variations as small as femtoteslas, roughly one-billionth the strength of Earth's magnetic field.[11] Medical applications include brain imaging that detects Alzheimer›s disease, epilepsy, and traumatic brain injury earlier than classical methods—potentially revolutionary for healthcare. The same sensors detect submarines at ranges beyond current sonar, track vehicles through magnetic signatures they cannot mask, and monitor neural activity from outside the skull.

Think of it like the difference between a security guard watching the front door and a guard who can sense your heartbeat through the walls. Both are surveillance. But one you can see and potentially evade. The other operates on physical principles most people don't understand and can't easily defend against.

Quantum positioning systems enable navigation without GPS by measuring acceleration and rotation with extreme precision using quantum sensors.[12] The beneficial applications are obvious: navigation when GPS is unavailable—underwater, indoors, in space—and backup systems for critical infrastructure. The concerning applications are equally clear: military systems immune to GPS jamming, tracking that doesn't rely on external signals that can be detected or blocked.

The dual-use pattern repeats across quantum sensing technologies. Quantum gravimeters that find underground water also detect hidden installations. Quantum clocks that synchronize communications also enable location tracking with meter-level precision. Quantum radar that improves air traffic control also detects stealth aircraft.

Current surveillance relies on cameras, microphones, GPS tracking, network monitoring—all detectable, all possible to evade with knowledge and effort. A person who knows they're being watched can avoid cameras, leave their phone behind, use encrypted communications.

Quantum surveillance detects magnetic signatures, gravitational anomalies, quantum states—harder to detect, harder to evade, operating on physical principles difficult to shield against. You cannot leave your magnetic signature behind. You cannot shield against gravitational measurements from aerial platforms. The surveillance becomes ambient, unavoidable, operating on physics rather than technology you can choose to use or not use.

The Regulation Gap

International law governing quantum sensor deployment does not exist. Privacy legislation written for cameras and microphones does not contemplate sensors that detect magnetic fields through walls or map gravitational anomalies from aircraft. The technology is advancing faster than any regulatory body has begun to consider.

By 2026, the transition from laboratory to deployment[10] is already happening. Medical imaging systems are in clinical trials. Military sensing programmes are in testing. Commercial geological applications are running. Once deployed at scale, restricting them becomes politically impossible.

Once deployed widely, restricting them becomes politically difficult. The logic is familiar: "Other nations use quantum sensors for security. We need them too. Our adversaries won't limit their capabilities, so we can't limit ours." The surveillance capabilities ratchet upward, one deployment at a time, one security justification at a time.

The Optimization Apocalypse

Quantum computers are exceptionally well-suited to optimization problems—finding the best solution across trillions of possibilities in logistics, finance, traffic routing, and resource allocation. The same capability that makes them valuable for corporations makes them dangerous for the people whose jobs those corporations are currently paying humans to do.

This creates value. It also destroys jobs.

Every previous wave of automation displaced one category of worker while creating new ones. Manufacturing gave way to service. Office work gave way to digital work. Each transition was painful and each created more jobs than it destroyed—eventually. Quantum optimization is different, and the difference matters.

The Scenario Nobody Wants to Plan For

Consider a hypothetical logistics company—call it LanePath Logistics, operating in a major European port city—that in 2019 employed 180 route planners. Each managed delivery networks for their region—millions of packages annually, thousands of variables, routes optimised through experience and classical algorithms honed over careers.

In 2027—this is a plausible scenario, not a documented case—the company deployed a quantum-classical hybrid optimization system. The algorithm analyzed 47 million route combinations per second, finding paths 23% more efficient than human planners achieved. Deliveries arrived faster. Fuel consumption dropped. Costs fell.

Over eighteen months, the planning staff shrank from 180 to 52. The displaced workers—logistics managers with 10–20 years of experience—discovered that their expertise had become computationally obsolete. The knowledge they'd built over careers could be replicated by quantum circuits running optimization algorithms humans hadn't designed.[13]

Imagine a scenario like this. The industries are different but the economic logic is identical:

Financial services: In financial services, quantum algorithms optimising portfolios and risk models[14] will not eliminate the jobs entirely—they will consolidate them in organisations with quantum access. A firm without quantum capabilities will produce good analysis. A firm with quantum will produce mathematically optimal analysis. The difference compounds, and it is not survivable for the firm that falls behind

Manufacturing: In manufacturing, quantum optimisation of production schedules and inventory systems creates efficiency advantages that market pressure forces competitors to match. Workers whose roles centred on optimisation—production planners, inventory managers, logistics coordinators—find their positions automated not because their employers chose automation but because their competitors forced the issue.

The Inequality Amplifier

Technology usually creates jobs while destroying others. The internet destroyed video rental stores but created e-commerce, social media, streaming platforms, app development—millions of new jobs globally.

This isn't like that.

The jobs quantum optimization creates—quantum algorithm developers, quantum hardware engineers, quantum-AI researchers—require PhD-level expertise in quantum mechanics, mathematics, and computer science. The jobs it destroys—logistics managers, financial analysts, production schedulers—require different skills built over careers in those industries.

Retraining a logistics manager to become a quantum algorithm developer is not a realistic pathway. It is the kind of thing that sounds reasonable in a government white paper and fails comprehensively in practice—the skill gap is measured in years of specialist training, not months of vocational retraining.

The Timeline and the Policy Gap

Quantum optimization applications are arriving 2027–2030. Job displacement could begin 2028–2032, accelerating through the 2030s as quantum computing capabilities improve and deployment costs fall.

That's 2–6 years before significant disruption begins. Are displaced workers being retrained? Are social safety nets being strengthened? Are industries developing transition plans?

Not visibly. Not at scale. Not with the urgency the timeline suggests.

The Quantum Divide

Quantum computers are expensive in a way that concentrates ownership. Building one requires specialised fabrication facilities that cost hundreds of millions of dollars, dilution refrigerators running $500,000 to $2 million each, teams of physicists and engineers with expertise measured in PhDs per department, years of R&D before anything computable emerges, and then continuous investment in maintenance, cooling, and calibration just to keep it running. The result is not a tool. It is infrastructure—the kind that only institutions with sovereign resources or corporate scale can build.

This creates stratification: quantum haves and quantum have-nots.

The Haves:

The haves are easy to enumerate: the United States, China, EU member states, the United Kingdom, Japan, Canada, and Australia—nations with combined quantum investment exceeding $80 billion. Plus the corporations: Google, IBM, Microsoft, Amazon, IonQ, Rigetti, PsiQuantum, Quantinuum, and Atom Computing, with billions in private capital deployed or committed.

The Have-Nots:

The have-nots are harder to see in quantum discussions, which is part of the problem. Most of Africa, Latin America, Southeast Asia, and the Middle East have research programmes, but nothing approaching competitive quantum development. Small businesses and startups in quantum-dependent nations have no quantum access except through foreign-owned cloud platforms. The dependency structure is already forming.

The Dependencies This Creates

This mirrors historical patterns of technological dependency.

In the 1970s, OPEC nations controlled oil production and could dictate terms to dependent economies. Oil-importing nations faced supply disruptions, price shocks, and strategic vulnerability. The dependency shaped foreign policy, military strategy, and economic development for decades.

In the 2020s, a handful of companies control advanced semiconductor fabrication. Taiwan Semiconductor Manufacturing Company, Samsung, and Intel produce the cutting-edge chips that power everything from smartphones to weapons systems. Nations without semiconductor fabs depend on these suppliers for technology essential to economic competitiveness and national security. When chip shortages disrupted global supply chains in 2021–2023, automobile manufacturers, consumer electronics companies, and defense contractors faced production halts—because they depended on components they couldn't produce domestically.

Quantum computing is following the same pattern. But quantum dependency runs deeper than oil or semiconductors.

Oil powers vehicles and generates electricity—crucial, but substitutable over time through renewable energy and electric vehicles. Semiconductors enable computing and communications—essential, but producible if nations invest in fabrication capacity.

Quantum computers don't just power devices or enable communications. They determine what's cryptographically secure, what molecules are discoverable, what materials are designable, what systems are optimizable. A nation without quantum capabilities faces cryptographic vulnerability, pharmaceutical disadvantage, materials science gap, and optimization deficit simultaneously.

Quantum-dependent nations face a specific irony:

They will have to buy protection against quantum threats from the same nations whose quantum programmes create those threats.

They pay for drugs and materials designed on quantum computers they cannot build.

They run their AI on cloud platforms owned by foreign corporations, under foreign legal jurisdiction.

And they pay ongoing access fees to quantum-capable nations for the privilege of competing economically against industries those same nations have already optimised.

The Strategic Implications

Without deliberate intervention, the quantum divide hardens into something that looks less like a technology gap and more like colonial dependency—nations permanently positioned as consumers of quantum technology produced elsewhere, unable to develop domestic cryptographic security, competing against quantum-optimised industries they cannot match, and purchasing the outputs of a civilisational capability they had no part in building

The Technology Transfer Dilemma

The dual-use problem, described in Chapter 17, runs through every quantum technology: the same physics that heals also spies, and the same hardware that discovers drugs also simulates weapons.

This makes international cooperation complicated. Export controls restrict quantum technology sales to adversarial nations.[15] But restrictions fragment the quantum ecosystem, creating incompatible standards and reducing global scientific collaboration.

The likely outcome: quantum-capable nations form alliances—quantum technology sharing agreements similar to nuclear cooperation frameworks. Quantum-dependent nations negotiate access through diplomatic and economic concessions. Quantum-excluded nations fall further behind, facing widening technological gaps that compound over decades.

The Inequality Cascade

Small quantum advantages compound in ways that look modest until they don't.

Year 1: A 5% efficiency improvement from optimisation in year one looks like noise.

Year 5: the same continuous improvement has produced a 25% cumulative advantage—competitors are struggling to match pricing.

Year 10: By year ten, the gap is structural: the quantum-enabled industry dominates its market, attracts capital, and funds further quantum research. The firms without quantum access cannot catch up through incremental improvement. The gap is too large. And it was built from 5% at a time.

The Alternative Vision

Quantum computing could instead be treated as a global public good—international research centres with multilateral governance, open-source quantum software, subsidised cloud access for researchers in developing nations, coordinated education programmes.

This requires political will and international cooperation. The current trajectory suggests stratification instead.

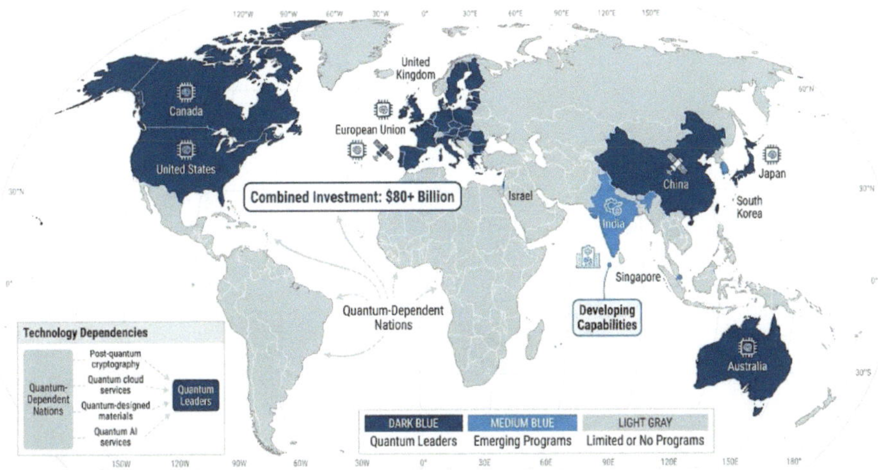

The Quantum Divide: 85% of global investment concentrated in wealthy nations. Most of the world depends on quantum services from quantum-capable countries.

The Environmental Paradox

Here's the irony that keeps environmental physicists up at night.

The technology we're building to solve climate change—by discovering better batteries, designing greener catalysts, optimising energy grids—currently consumes approximately the same electricity as 20 to 80 average homes per system. Continuously. Just to maintain the temperature required for qubits to function.

Current quantum computers require cooling to 10–20 millikelvin—colder than outer space, colder than the cosmic microwave background radiation left over from the Big Bang.[16] These aren't metaphorical comparisons. Outer space averages 2.7 Kelvin. Quantum processors operate at 0.01–0.02 Kelvin—more than 100 times colder than the vacuum of space.

Each dilution refrigerator drawing its 25–100 kilowatts. Every day. Every hour. To keep a small cylinder a fraction of a degree above absolute zero, so that a few hundred fragile qubits can perform calculations for a few microseconds before decoherence takes them.

If quantum computing stays where it is—expensive, specialised, deployed in small numbers at research facilities and national labs—the energy footprint is negligible. But quantum computing is not going to stay there. If it follows the trajectory of classical computing—exponential growth in deployment as it becomes useful and accessible—the aggregate energy consumption will be significant.

One system consuming 25–100 kilowatts is a curiosity. Ten thousand systems consuming that amount continuously is a power grid problem.

And then there's helium-3.

The Helium-3 Constraint

Dilution refrigerators—the devices that keep quantum computers cold—require helium-3, an isotope so rare that global production runs to approximately 8,000 litres per year.[17] To put that in perspective: 8,000 litres is roughly four average bathtubs.

The world's entire annual helium-3 production would fill four bathtubs.

Helium-3 is produced primarily as a byproduct of tritium decay in nuclear weapons facilities. You cannot simply make more of it by building new factories. The supply is fundamentally constrained by how many nuclear warheads exist and how long they've been sitting around decaying. America has some. Russia has more. The total global supply is small, declining as warheads are dismantled, and currently allocated between scientific research, medical imaging (MRI machines use helium-3), and quantum computers.

As quantum computers scale from dozens of research systems to thousands of commercial deployments, demand for helium-3 will increase substantially. The supply cannot grow to meet it. At some point, the quantum industry will be competing with hospital MRI machines for a resource that neither can substitute.

This is the helium-3 constraint—a physical resource limitation on quantum scaling that appears in almost no quantum computing coverage because it doesn't fit the narrative of inevitable, frictionless progress. It's the detail that gets left out of the roadmaps, the investor decks, and the government strategy documents. Not because it's a secret, but because it's uncomfortable.

Solutions exist in principle: more efficient refrigerator designs, alternative cooling approaches, helium-4 dilution systems that don't require helium-3, and—most interestingly—photonic or room-temperature quantum approaches that don't need refrigeration at all. Whether those solutions mature fast enough to meet scaling demand is an open question.

The Aviation Parallel

Think of it like the early days of powered flight. Aviation transformed civilization, enabling global travel, international commerce, and cultural exchange impossible before. The environmental cost of jet fuel wasn't a consideration in 1910 when the industry was getting started. By the time aviation scaled globally, addressing the environmental impact required retrofitting an entire industry—new engines, new fuels, new regulations, all fighting against entrenched infrastructure and economic incentives.

We have the chance to design quantum computing for energy efficiency from the beginning—before the infrastructure becomes entrenched, before thousands of systems are deployed, before economic incentives lock in energy-intensive architectures. Prioritizing energy-efficient quantum designs, requiring renewable energy for quantum facilities, restricting quantum deployment to high-value applications where benefits exceed environmental costs.

The current trajectory suggests we're not taking it. Market forces are driving deployment. Energy efficiency is a secondary concern. The environmental consequences will be addressed reactively, if at all, after the industry is established.

The Weaponization Question

Quantum technology has military applications that create strategic instability.

The most immediate: breaking encryption while protecting your own. The nation that achieves quantum RSA-breaking capability first—without adversaries knowing—gains asymmetric intelligence advantage. That means reading adversary military communications, accessing classified networks, and decrypting diplomatic cables from the past decade. The advantage persists until the adversary catches up or migrates to quantum-resistant cryptography—whichever comes first.

The race dynamic is simple and inevitable: break adversary encryption while protecting your own. Every nation with quantum programmes understands this. Every intelligence agency with quantum funding is pursuing it. The first to achieve breaking capability without announcing it gains intelligence advantage measured in years.

The Good News: The Replacement Already Exists

After several pages of escalating cryptographic dread, I owe you something: the solution exists. It's not perfect. It requires effort. But it's real, it's standardised, and it works.

In August 2024, the National Institute of Standards and Technology finalised three post-quantum cryptographic algorithms—mathematical structures that can resist attacks from quantum computers.[20] They have names that only a cryptographer could love:

CRYSTALS-Kyber (now officially called ML-KEM)—for key encapsulation. This is the mechanism that establishes a shared secret between two computers. Currently, that's done using mathematical problems that quantum computers can break. Kyber replaces it with lattice-based cryptography—a different class of mathematical problem that quantum algorithms cannot efficiently solve.[21]

CRYSTALS-Dilithium (now ML-DSA)—for digital signatures. When your browser displays the padlock icon, it's verifying that you're actually talking to your bank and not an imposter. That verification uses digital signatures. Dilithium replaces the current signature scheme with one that quantum computers cannot forge.

SPHINCS+ (now SLH-DSA)—a backup signature scheme using hash-based cryptography, included because diverse approaches make the overall system more robust.

Here's how to think about what "lattice-based cryptography" means without spending a year studying mathematics:

Imagine a three-dimensional city of skyscrapers. Classical cryptography is like hiding something in a specific apartment in one building—if you're clever enough, you can figure out which building, which floor, which door. Lattice-based cryptography is like hiding something in an apartment in a city with *millions of dimensions*—not three dimensions, but millions. Even a quantum computer gets lost trying to navigate a mathematical space with millions of dimensions. The hiding place is simply too well-hidden.

The problem is not the mathematics. The mathematics is solid. Multiple independent teams have verified it. NIST took eight years and ran an open international competition—essentially crowdsourcing verification to the world's best cryptographers—before finalising these standards. The algorithms are ready.

The problem is the migration.

What Actually Needs to Change

Every system that uses RSA or elliptic curve cryptography—which is essentially every system that communicates securely over the internet—needs to be updated. This means:

Your bank's authentication system. Your government's secure communications. Hospital records systems. Military encryption. Smart grid controllers. Satellite communication. SWIFT, the financial messaging network connecting every bank on earth. TLS (the "S" in HTTPS). SSH. PGP. VPNs. Password managers. Messaging apps.

Some of these are easy to update—a software patch, a configuration change, a library update. Some require hardware replacement—older systems with cryptographic chips that cannot be reprogrammed. Some require coordination across thousands of institutions with no central authority to mandate the change. And all of it needs to happen *before* a quantum computer powerful enough to break RSA comes online. Migration at the back of an existential deadline.

As of 2026, migration is underway in the United States federal government (mandated by NSM-10, signed 2022), in the financial sector (SWIFT began testing PQC in 2024), and in cloud providers (Google, Cloudflare, and Amazon have implemented PQC in their TLS stacks). The roughly 60% of critical systems that haven't begun migration are mostly legacy systems in healthcare, manufacturing, and government infrastructure—exactly the systems with the longest update cycles and the hardest migration paths.

What You Can Actually Do

For individuals: your immediate exposure is limited. Personal password managers and messaging apps (Signal implemented PQC in 2024) are updating. The risk is not your personal communications—it's the institutions holding your data.

Ask your bank whether they have a post-quantum cryptography migration plan. If they don't know what that means, that's a meaningful data point about how seriously they're taking it. Ask your company's IT team the same question. The correct answer is "we're working on it." The concerning answer is "we'll deal with that when quantum computers arrive."

The 2030s will arrive whether or not you're ready for them.

Quantum Sensing for Military Applications

Quantum magnetometers can detect submarines at greater range than current sonar by measuring minute magnetic field disturbances caused by steel hulls, electrical systems, and propulsion.[18] This doesn't just improve existing capabilities—it changes naval warfare doctrine. Submarines are effective because they're difficult to detect. Quantum sensors make them detectable at distances that eliminate stealth advantages.

Quantum gravimeters can map underground facilities from aerial or satellite platforms. The same sensors that found Birmingham's lost tunnel can detect bunkers, command centres, weapons storage facilities, underground transportation networks. Military installations designed to be hidden become visible through gravitational mapping.

Quantum inertial sensors enable navigation without GPS—measuring acceleration and rotation with quantum precision, immune to jamming, untraceable by adversaries monitoring GPS signals. Military systems that don't rely on external signals that can be jammed or spoofed gain operational advantages in contested environments.

The dual-use nature of these technologies is well-established in the published scientific literature.[18] Defence research agencies in multiple countries have publicly acknowledged quantum sensing programmes. The gap between laboratory demonstration and operational military deployment remains significant; this discussion reflects publicly available research, not classified capability assessments.

Quantum Communication Networks

Quantum key distribution (QKD) creates theoretically unbreakable military communications using quantum mechanics rather than mathematical complexity. China has deployed satellite-based QKD systems connecting military and government facilities.[19]

The advantage: secure communications immune to quantum computers breaking encryption. The disadvantage: requires specialized infrastructure—fiber networks or satellite systems—not universally deployable.

The result: quantum communication advantages accrue to nations with resources to build quantum satellite networks and fiber infrastructure. Smaller nations and non-state actors cannot afford equivalent capabilities.

The Escalation Dynamics

The structure creates escalation pressure:

Quantum capabilities create military advantage. Advantage creates pressure for adversaries to match or exceed. Matching requires investment in quantum programmes. Investment accelerates development. Development creates new capabilities. New capabilities create new advantages. The cycle continues.

This is familiar from nuclear weapons, ballistic missiles, stealth aircraft, cyber capabilities. Each technology that provides military advantage triggers arms race dynamics. Quantum technology is following the pattern.

The Control Problem

Nuclear weapons have non-proliferation treaties, export controls, international monitoring. The frameworks are imperfect—North Korea, Pakistan, and India developed nuclear weapons despite non-proliferation efforts. But the frameworks exist. They create diplomatic pressure, impose economic costs on proliferation, enable verification and inspections.

Quantum technology has no equivalent. Quantum computers are dual-use—civilian and military applications are intertwined and technically inseparable. A quantum computer that simulates pharmaceutical molecules also simulates weapons materials. Quantum sensors for medical imaging also detect submarines. You cannot restrict quantum research without restricting scientific progress.

Export controls limit quantum hardware sales to adversarial nations.[15] But quantum algorithms are published in academic journals. Quantum software is open-source. The knowledge proliferates even when the hardware doesn't.

The result is limited control over proliferation, and no credible mechanism for building more. How do you prevent a quantum arms race when the physics is published in open journals, the algorithms are openly available, and even the hardware is becoming commercially accessible? Nobody has a good answer. The frameworks that worked imperfectly for nuclear weapons—physical size, scarce materials, visible infrastructure—do not translate to quantum. A quantum algorithm fits in a text file. You cannot inspect it from a satellite

No good answers yet. We're building quantum capabilities without frameworks to prevent destabilization.

The Governance Gap

We're building transformative technology without governance structures to manage it.

What's Missing:

International quantum standards: No global agreement on quantum communication protocols, quantum computing benchmarks, or quantum sensor specifications. Different nations are developing incompatible systems that won't interoperate.

Export control frameworks: Patchwork restrictions on quantum technology sales. Some nations restrict exports to adversaries, others don't. No coordinated international approach.[15]

Privacy regulations addressing quantum sensing: Laws written for cameras and microphones don't contemplate quantum sensors detecting magnetic signatures, gravitational anomalies, or quantum states.

Quantum surveillance capabilities are advancing faster than privacy law.

Ethical frameworks for quantum AI: No consensus on acceptable uses of quantum-enhanced AI for surveillance, optimization, or decision-making affecting human lives.

Post-quantum cryptography migration mandates: Some nations require government systems to migrate to quantum-resistant cryptography. Most don't. Private sector migration is largely voluntary, proceeding slower than threat timeline suggests necessary.

Quantum workforce development coordination: Fragmented university programmes, no international collaboration on quantum education, workforce shortages worsening as quantum programmes expand.

Environmental regulations for quantum computing: No standards for quantum computer energy efficiency, cooling system sustainability, or helium-3 usage management.

The Pattern

Governance lags technology. This is historical pattern, not unique to quantum computing:

Automobiles existed for decades before seatbelt laws, emissions standards, or safety regulations. Early adoption was unregulated. Deaths accumulated. Regulations came later, reactively, after the industry was established and resistant to requirements.

The internet existed for decades before privacy regulations, content moderation frameworks, or cybersecurity standards. Early development prioritized openness and growth. Security and privacy were afterthoughts. We're still dealing with consequences.

Social media platforms reached billions of users before content moderation policies, misinformation controls, or privacy protections.

The platforms were designed without governance in mind. Regulations are being developed now, reactively, against entrenched economic interests and established user expectations.

We're repeating this pattern with quantum computing. Development first, governance later, consequences addressed reactively when they become unavoidable.

The Coordination Problem

Effective quantum governance requires:

International cooperation during a period of geopolitical tension and strategic competition. Nations view quantum technology as competitive advantage, not shared challenge requiring coordination.

Technical expertise among policymakers who often lack quantum physics or computer science backgrounds. The gap between technical complexity and policy understanding creates governance failures.

Anticipatory regulation—creating rules for technology still under development. Difficult because capabilities are uncertain and potential impacts are speculative.

Balancing innovation and control. Overregulation stifles scientific progress and economic development. Underregulation creates risks that become crises.

None of this is easy. All of it is necessary.

The Current State

Some nations are developing quantum strategies—China's national quantum programme, the U.S. National Quantum Initiative, the EU Quantum Flagship. Some corporations are establishing quantum ethics boards—IBM, Google, Microsoft have internal guidelines for quantum research. Some researchers are publishing quantum governance frameworks in academic journals.

But comprehensive, coordinated, international quantum governance?

Doesn't exist.

We're building quantum capabilities faster than we're building governance capacity.

The Window Is Closing

Quantum computers capable of breaking RSA-2048 are potentially 4–6 years away.

Establishing the governance frameworks, completing the cryptographic migration, developing the privacy regulations, creating the workforce programmes, and implementing the environmental standards would take 5–10 years under optimistic assumptions.

The capabilities are arriving faster than the governance structures.

This isn't inevitable. Nuclear technology arrived in 1945 with the Trinity test and Hiroshima. The Nuclear Non-Proliferation Treaty was signed in 1968—twenty-three years of development before comprehensive international governance. But those twenty-three years included close calls that nearly ended civilization: the Cuban Missile Crisis, nuclear standoffs, arms races that consumed enormous resources and created global instability.

We could learn from that history. Establish quantum governance frameworks now, before capabilities outrun control. Coordinate international standards. Mandate post-quantum cryptography migration. Develop privacy protections for quantum sensing. Create inclusive access policies to prevent quantum divide from hardening into permanent inequality. Implement environmental standards before energy-intensive infrastructure becomes entrenched.

Or we could repeat the pattern: build first, govern later, deal with consequences as they emerge through crisis management and reactive policymaking.

Current trajectory suggests the latter.

What Happens If We Get It Wrong

The optimistic scenario: quantum computing transforms industries, accelerates scientific discovery, solves intractable problems in materials science and drug development. Benefits accrue broadly through commercial deployment, international cooperation, and inclusive access policies.

The pessimistic scenario: quantum computing breaks encryption before critical infrastructure is protected, enables mass surveillance without privacy safeguards, widens economic inequality between quantum-capable and quantum-dependent nations, and creates strategic instability through quantum arms races. Benefits concentrate in quantum-capable nations and corporations while most of the world faces disadvantage.

The realistic scenario: probably somewhere between these extremes. Some benefits realized. Some risks manifested. Messy transition with winners and losers, requiring reactive policy responses and damage control.

The Plausible Failure Modes:

Scenario 1: The Y2Q Crisis Quantum computers achieve RSA-breaking capability in 2030–2031. Post-quantum cryptography migration is 40% complete—better than 2026 levels but insufficient. Critical infrastructure remains vulnerable during the transition window.

Adversarial nations exploit this window. Intelligence services read encrypted military communications from the past five years, accessing operational plans, intelligence sources, diplomatic strategies. Financial institutions face emergency security updates as banking authentication systems become compromised. Corporations discover that encrypted IP shared with partners is now readable to competitors with quantum access.

The scramble to complete migration creates chaos. Legacy systems that can't be updated require replacement. Supply chain disruptions as hardware vendors struggle to meet sudden demand for quantum-resistant components. Organizations discover that quantum-resistant cryptography breaks compatibility with systems designed around RSA assumptions.

Likelihood: This scenario is plausible if quantum hardware advances faster than cryptographic migration—which current timelines suggest is possible. The 40% migration rate is optimistic; many organizations haven't begun planning. The window of vulnerability could be wider than estimated.

Scenario 2: Surveillance Expansion Quantum sensors deployed initially for legitimate purposes—medical imaging, geological survey, infrastructure monitoring—become surveillance infrastructure. The technology that finds underground water also maps hidden facilities. The sensors that detect early-stage disease also monitor populations.

Authoritarian regimes deploy quantum surveillance extensively—gravitational mapping of urban areas to track mass movements, magnetic sensors monitoring gatherings, quantum positioning systems tracking individuals through physical signatures they cannot mask. Democratic nations face pressure to deploy equivalent capabilities for security purposes. "Our adversaries use quantum sensors. We need them too for public safety."

Privacy becomes impossible against quantum sensing. Classical privacy protections—avoiding cameras, leaving devices behind, using encryption—become ineffective against sensors detecting unavoidable physical signatures. The surveillance capabilities ratchet upward globally, one security justification at a time.

Likelihood: Already beginning. Quantum sensors are transitioning from laboratory to operational deployment in 2026. The dual-use nature makes this trajectory difficult to prevent without restricting beneficial applications. Medical quantum imaging and surveillance quantum mapping use identical physics.

Scenario 3: The Quantum Divide Hardens Quantum computing capabilities concentrate in wealthy nations and large corporations. The initial quantum advantage—5% efficiency improvements from optimization, better materials discovered through simulation, pharmaceutical development accelerated by quantum chemistry—seems modest.

But advantages compound. Industries with quantum access optimise continuously, developing better processes, attracting more capital, funding further quantum research. Industries without quantum access fall progressively behind. Small gaps in Year 1 become structural disadvantages by Year 10.

Quantum-dependent nations become technologically subordinate—dependent on foreign quantum cloud services, unable to develop quantum-resistant cryptography domestically, competing against quantum-optimised industries without quantum access. The economic inequality between quantum-capable and quantum-dependent nations widens over decades, creating political instability and resentment similar to historical resource dependencies.

Likelihood: This is the current trajectory without deliberate intervention. Market forces concentrate quantum capabilities in organizations with resources to invest. International cooperation on inclusive quantum access is limited. Export controls prevent

technology transfer to adversarial nations, but also limit access for developing nations.

Scenario 4: Workforce Displacement Without Transition Support Quantum optimization displaces workers in logistics, finance, manufacturing, and supply chain management starting in 2028–2030. The displacement happens faster than workforce development programmes can respond.

Logistics managers with 20 years of experience discover their expertise has become computationally obsolete. Financial analysts find their roles automated by quantum-optimised risk models.

Production schedulers are replaced by quantum optimization systems that plan more efficiently than humans.

Retraining programmes prove insufficient. The new jobs—quantum algorithm developers, quantum hardware engineers—require PhD-level expertise. The displaced workers need income within months, not the years required to retrain for quantum roles. Social safety nets designed for temporary unemployment face sustained displacement.

Economic disruption creates political instability in affected regions.

Populist movements emerge opposing quantum technology deployment.

Governments face pressure to restrict quantum optimization to protect jobs, conflicting with economic pressure to remain competitive with quantum-adopting nations.

Likelihood: Depends on deployment speed and policy response.

Historical precedent suggests technology-driven displacement creates painful transitions even when long-term outcomes are positive. Coal mining to renewable energy, manufacturing to services, knowledge work to AI-augmented work—all created displacement requiring decade-scale adjustments.

Scenario 5: Environmental Costs Exceed Benefits Quantum computing scales to thousands of installations globally following classical computing's exponential growth trajectory rather than remaining specialized. Aggregate energy consumption reaches significant levels—gigawatts for quantum infrastructure globally.

Helium-3 supplies face severe constraints. Dilution refrigerator costs increase as scarcity drives prices up. Alternative cooling approaches remain unproven at scale. Room-temperature quantum systems don't materialize on timelines industry projected.

The environmental costs of quantum infrastructure—continuous high energy draw, scarce resource consumption, cooling system emissions—exceed the climate benefits from quantum-enabled discoveries. Carbon capture catalysts designed with quantum computers prove less impactful than anticipated. Better batteries help but don't offset quantum infrastructure environmental footprint.

Likelihood: Least likely scenario if quantum computing remains specialized for specific high-value applications rather than becoming ubiquitous general-purpose technology. But possible if industry follows classical computing's path toward widespread deployment without environmental constraints.

These scenarios aren't mutually exclusive. They could occur

simultaneously: Y2Q crisis creates security instability, which accelerates quantum surveillance deployment, while quantum divide widens and job displacement creates economic stress. Compounding risks create systemic fragility.

The Alternative

We could choose differently:

Aggressive post-quantum cryptography migration with government mandates and industry coordination—preventing Y2Q crisis.

Privacy regulations for quantum sensing established before widespread deployment—limiting surveillance expansion.

International cooperation on quantum access through subsidized cloud services, open-source quantum software, and multilateral research centres—reducing quantum divide.

Workforce retraining programmes and strengthened social safety nets started now, before displacement begins—mitigating job displacement impacts.

Energy-efficient quantum architectures prioritized through regulation and funding incentives—minimizing environmental impact.

This requires recognizing risks early, prioritizing governance over pure capability development, and accepting that technology advancement without deliberate management creates predictable problems.

Closing

Quantum computing is not inherently good or bad. It's a tool. Like nuclear fission, genetic engineering, or artificial intelligence—technologies whose consequences depend on how we develop and deploy them.

Right now, we're developing quantum capabilities faster than governance frameworks. We're building encryption-breaking computers before migrating critical infrastructure to quantum-resistant cryptography. We're deploying quantum sensors before establishing privacy protections. We're optimizing systems before preparing for workforce displacement. We're creating quantum divide without policies addressing technological inequality.

This is predictably problematic.

The risks aren't hypothetical. They're emerging now:

Adversaries are harvesting encrypted traffic for future decryption—confirmed by security researchers and intelligence officials.

Quantum sensors are transitioning from laboratories to operational deployment in medical, military, and commercial applications.

Quantum optimization is beginning to transform logistics, finance, and manufacturing—with job displacement impacts on the horizon.

Quantum capabilities are concentrating in wealthy nations and large corporations—creating dependencies similar to oil and semiconductors.

International governance frameworks are lagging technology development—fragmented, incomplete, insufficient for the challenges ahead.

We have 5–10 years before quantum computers become powerful enough to create these problems at scale. That's enough time to prepare—if we start now.

The tools exist. Post-quantum cryptography works and is ready to deploy. Privacy frameworks can be written before surveillance capabilities are entrenched. Workforce transition programmes are standard policy—they require political will and budget, not invention. Environmental standards for energy-intensive industries are not new. Inclusive access policies have been built for previous technologies. The question is not capability. It is urgency.

But this requires urgency most governments and organizations haven't demonstrated.

The question isn't whether quantum computing transforms society. It will. The question is whether we shape that transformation deliberately—through governance, planning, and international cooperation—or let it shape us through market forces, geopolitical competition, and reactive crisis management.

So far, we're choosing the latter.

The quantum revolution is coming whether we're ready or not. The window for deliberate choice is closing. Once quantum computers break encryption at scale, quantum sensors deploy widely, and quantum optimization displaces workers in millions—the trajectory becomes much harder to change.

We need to get this right. And we're running out of time.

Turn the page.

References

1 Mosca, M. (2018). Cybersecurity in the quantum era: Prepare now or pay later. IEEE Security & Privacy, 16(5), 38–41.

2 Author›s estimate based on NIST (2024) post-quantum cryptography implementation guidance and industry adoption patterns for previous cryptographic standard transitions.

3 Lenstra, A. K., et al. (2001). Selecting cryptographic key sizes. Journal of Cryptology, 14, 255–293.

4 Shor, P. W. (1994). Algorithms for quantum computation: Discrete logarithms and factoring. Proceedings 35th Annual Symposium on Foundations of Computer Science, 124–134.

[5] National Academies of Sciences, Engineering, and Medicine (2019). Quantum Computing: Progress and Prospects. National Academies Press.

[6] National Institute of Standards and Technology (2024). Post-Quantum Cryptography Standardization. Federal Information Processing Standards Publication.

[7] Grover, L. K. (1996). A fast quantum mechanical algorithm for database search. Proceedings 28th Annual ACM Symposium on Theory of Computing, 212–219.

[8] Aggarwal, D., et al. (2018). Quantum attacks on Bitcoin, and how to protect against them. Ledger, 3, 68–90.

[9] Author's assessment based on public reporting by security researchers and intelligence community statements regarding adversarial encrypted traffic collection programmes.

[10] Degen, C. L., Reinhard, F., & Cappellaro, P. (2017). Quantum sensing. Reviews of Modern Physics, 89, 035002.

[11] Budker, D., & Romalis, M. (2007). Optical magnetometry. Nature Physics, 3, 227–234.

[12] Geiger, R., et al. (2020). Detecting inertial effects with airborne matter-wave interferometry. Nature Communications, 11, 2611.

[13] Farhi, E., Goldstone, J., & Gutmann, S. (2014). A quantum approximate optimization algorithm. arXiv:1411.4028.

[14] Orus, R., Mugel, S., & Lizaso, E. (2019). Quantum computing for finance: Overview and prospects. Reviews in Physics, 4, 100028.

[15] U.S. Department of Commerce (2024). Export Administration Regulations: Quantum Computing Controls. Bureau of Industry and Security, Federal Register.

[16] Krinner, S., et al. (2019). Realizing repeated quantum error correction in a distance-three surface code. Nature, 605, 669–674.

[17] Shea, D. A., & Morgan, D. (2010). The helium-3 shortage: Supply, demand, and options for Congress. Congressional Research Service Report R41419.

[18] Author›s synthesis based on unclassified defense research publications regarding quantum sensing applications for submarine detection and underground facility mapping.

[19] Liao, S. K., et al. (2017). Satellite-to-ground quantum key distribution. Nature, 549, 43–47.

[20] NIST (2024). Federal Information Processing Standards 203, 204, and 205: Module-Lattice-Based Key-Encapsulation Mechanism, Module-Lattice-Based Digital Signature, and Stateless Hash-Based Digital Signature Standard. National Institute of Standards and Technology, August 13, 2024.

[21] Google Security Blog (2024). Protecting Chrome Traffic with Hybrid Kyber KEM. August 2023, with full deployment in Chrome 116.

[22] Signal (2024). PQXDH: Post-Quantum Extended Diffie-Hellman. Signal Technical Documentation, September 2023.

CHAPTER

20

Welcome to the Quantum Age

What twenty chapters of quantum physics actually means for the next twenty years of your life

You Were Quantum Before You Knew the Word

You are quantum. Not metaphorically. Literally.

The proteins in your retina use quantum coherence to detect single photons—converting a single particle of light into an electrical nerve signal with an efficiency that no human-designed detector has matched.[1] Your nose may detect the difference between molecules not by their shape (as was assumed for decades) but by their quantum vibrational signature—the specific frequencies at which their chemical bonds vibrate, detectable through a quantum mechanical process called inelastic electron tunneling.[2] The enzyme in your cells that helps convert food to energy—ATP synthase[14]—uses quantum tunneling to move protons across a membrane at rates that classical physics says should be impossible at body temperature.

And then there are the birds.

European robins navigate using Earth's magnetic field with a precision that has puzzled physicists for decades. The current leading explanation involves quantum entanglement inside specialized proteins in their eyes—cryptochrome proteins that create entangled electron pairs when struck by blue light. The orientation of the Earth's magnetic field influences how quickly those entangled electrons recombine, producing a signal the bird's nervous system uses as a compass.[3] A living creature navigating by quantum entanglement, without knowing what quantum entanglement is, without any of the dilution refrigerators and vacuum chambers we use to keep our qubits alive. Evolution stumbled onto quantum biology 300 million years before we invented the word.

Photosynthesis is the oldest quantum technology on Earth. When a plant captures sunlight, energy travels through the chlorophyll molecules using quantum coherence—the quantum equivalent of exploring all possible paths simultaneously and finding the most efficient route to the reaction centre. The efficiency of photosynthesis (capturing nearly all the energy in absorbed photons) has been partially attributed to this quantum coherent energy transfer. We have spent decades trying to replicate it in solar cells. We're still not as efficient as a leaf.[4]

This is what I want you to carry away from this book, above everything else: quantum mechanics is not something we invented. It's something we discovered. The universe has been running on quantum principles since before the first star formed.

> *Life figured out how to exploit quantum mechanics 3.5 billion years ago, using nothing but evolutionary time and chemical experimentation. We are late to the party.*

Extraordinarily, dangerously, excitingly late.

The quantum computers we are building are, at their core, an attempt to do deliberately and precisely what evolution has been doing accidentally for billions of years: exploit quantum mechanics to solve problems that are too hard for classical physics to handle.

> *We are trying to build what the mitochondria already are.*

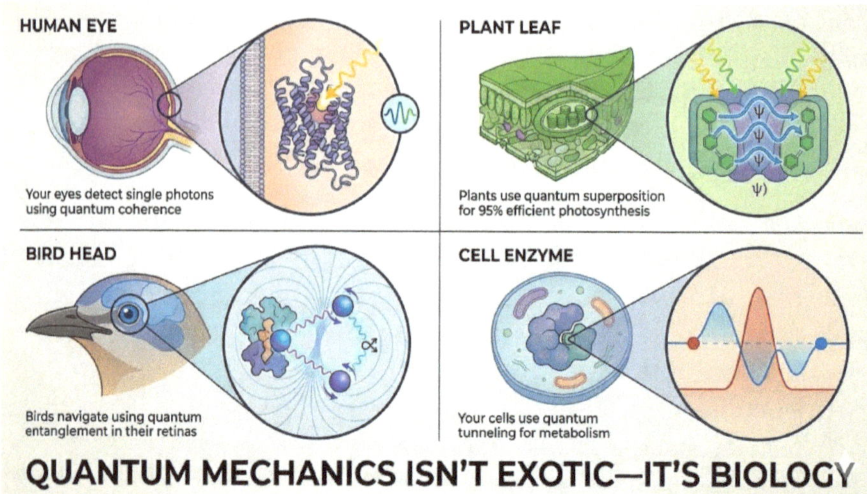

This matters because people think quantum computing is unnatural—forcing particles into strange states they wouldn't normally occupy. They think quantum computers are artificial constructs wrestling physics into submission.

Wrong.

Classical computers are unnatural. They force quantum particles to behave classically through brute-force engineering—massive error correction, continuous intervention, active suppression of quantum effects. Every transistor in your laptop is a quantum system being deliberately constrained to act classical.

> *We spent seventy years building technology that fights against quantum mechanics.*

The question isn't whether quantum mechanics works. It demonstrably works—it's running your cells right now.

The question is what we choose to do with it.

So. Twenty chapters of quantum physics. Let me tell you where we actually are.

Where Things Actually Stand

The honest picture is more specific than most quantum coverage suggests:

Quantum computers work. Not theoretically. Actually. Right now. In 2026.

They can simulate molecules with 20–60 atoms more accurately than classical computers.[5] A pharmaceutical company used quantum simulation to predict drug binding affinity 9% better than classical methods. That molecule is in pre-clinical testing. Nine percent doesn't sound revolutionary until you realize pharmaceutical companies test thousands of molecules to find one viable drug. Nine percent eliminates hundreds of failed experiments. Millions in costs. Months of time.

They can optimise logistics problems classical algorithms struggle with.[6] A European delivery company reduced route inefficiency by 7% using quantum optimization. That 7% saves €2.3 million annually in fuel costs. They're expanding the system.

They can detect patterns in data that classical machine learning misses.[7] A bank improved fraud detection by 5% using quantum kernels. Five percent means millions in prevented fraud.

They can discover better quantum error correction codes than humans designed in thirty years of trying.[8] Google's AI found three new codes optimised for different hardware. Those codes are being implemented in actual systems now.

What they can't do yet: Break RSA encryption. Simulate proteins at pharmaceutical scale. Run Grover's algorithm on databases large enough to matter. Solve general optimization better than classical supercomputers.

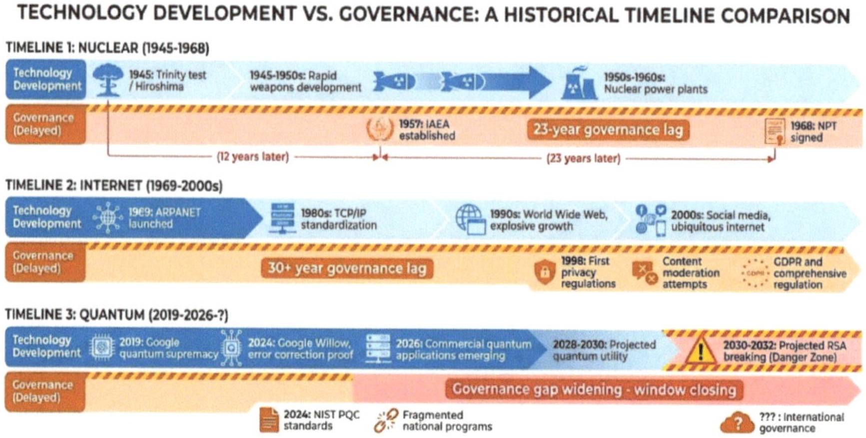

They work. They're expensive. They require expertise. And they're improving at a rate that makes the trajectory unmistakable, even when the arrival date isn't. The question is no longer whether. It's when.

The Cloud Changes Everything

You don't need to own a quantum computer to use one.

IBM Quantum Network provides cloud access to over 600 organizations[9] including a free tier that gives anyone with an email address quantum compute time on real hardware. You do not need $15 million or a physics department. You need a laptop.

Microsoft's Azure Quantum Elements targets chemistry and materials science[10] with several hundred enterprise customers by 2026 primarily in pharmaceutical and materials research. Amazon Braket connects users to hardware from multiple vendors[11] under standard cloud pricing.

The model is the same one that made cloud computing transformative: access without ownership. The same shift that let a two-person startup run enterprise-grade software on AWS is now letting pharmaceutical researchers run quantum chemistry on IBM's hardware. You do not need the machine. You need the connection.

The Market Reality

The global quantum computing market in 2026 runs to roughly $2 billion in annual revenue[12]—cloud platforms, hardware sales, software, consulting, and education combined

Compare to classical cloud computing: $600+ billion annually.[13] Quantum represents 0.3% of classical cloud. Which is a number that sounds negligible — until you remember what 0.3% of the internet looked like in 1994.

The size is less interesting than the trajectory. Growth rates of 40–60% annually, if sustained, put the quantum market at $20–30 billion by 2031—still small relative to classical computing, but no longer dismissible.

Most quantum companies are still burning cash faster than they generate revenue. Most applications are still in pilot programmes. The honest description of where the value sits in 2026 is: mostly future, partly present, and moving.

But it's past the question "will quantum computing work?" We're firmly in "when does it become practical for mainstream applications?"

Answer: Soon. Very soon.

The Next Five Years (Realistic Timeline)

Every quantum computing timeline published in the past ten years has been wrong in one direction or the other. Here is the honest range, based on current hardware trajectories rather than company roadmaps:

2027–2028: The Utility Threshold

Quantum chemistry integrated into standard pharmaceutical R&D pipelines, used routinely rather than experimentally. Quantum systems exceeding 10,000 physical qubits operational. Quantum optimization demonstrating clear commercial advantage at multiple major companies with documented results.

First quantum-classical hybrid AI models outperforming pure classical approaches on specific benchmarks. Global quantum cloud revenue exceeding $1.5 billion annually.

2029–2030: The Commercial Transition

Post-quantum cryptography mandated for government systems in major nations as quantum threat becomes concrete. First quantum-discovered drug candidate enters clinical trials. More than 100 corporations operating internal quantum teams.

Quantum advantage demonstrated across multiple distinct domains: chemistry, optimization, machine learning, materials, simulation. Quantum networking testbeds connecting cities beyond point-to-point links.

Y2Q migration deadline panic as organizations realize quantum encryption-breaking is imminent, triggering emergency cryptography upgrades.

2031 and Beyond: The Deployment Era

Quantum computers routinely used for drug discovery at major pharmaceutical companies as standard practice. Post-quantum cryptography migration 80%+ complete for critical infrastructure. Quantum advantage in materials discovery measurably accelerating battery, catalyst, and materials development.

Fault-tolerant quantum computing demonstrated at small scale—proof that error correction works as theorized, though not yet scaled to solve large practical problems.

What Could Go Wrong

These predictions rest on several assumptions that could fail. Hardware progress could slow or plateau in ways current trajectories do not predict. Error correction, which has looked increasingly tractable, could hit fundamental obstacles. Algorithms for problems that genuinely benefit from quantum advantage may prove harder to find than expected. Investment could dry up in a funding winter if milestones slip and patience runs out. If one assumption fails, add three to five years. If several fail simultaneously, the industry contracts, consolidates, and the timeline extends significantly. Quantum winters have happened before. They end.

What Could Go Better

The upside scenario also exists. These predictions assume incremental progress. Breakthroughs do not work that way. A novel qubit design that dramatically reduces error rates, an unexpected application that works with far fewer qubits than current estimates require, an error correction scheme that scales better than theoretical models predict—any of these could compress a decade into three years. The history of computing includes several such compressions. Assuming the trajectory will be smooth is itself an assumption.

The range runs from 2028–2030 in the optimistic scenario to 2032–2035 in the conservative one. Most likely outcome: messy, uneven, domain-specific progress—some applications arriving ahead of schedule, others years late, the overall picture looking nothing like any of the roadmaps.

In 2027 or 2028 — give or take a year — quantum chemistry becomes a routine tool in pharmaceutical R&D pipelines rather than an experimental one. Not because of a single breakthrough, but because error rates will have crossed the threshold where results are reliable enough to trust. Around the same time, quantum optimisation demonstrably outperforms classical approaches at two or three major corporations, with publicly documented results. Not press-release claims. Measured outcomes. That's the bellwether. If those milestones arrive on schedule, the optimistic timeline becomes credible. If they slip, add three years to everything.

Sometime between 2029 and 2031—and this one carries genuine urgency—post-quantum cryptography migration becomes mandatory for government systems in major nations because the quantum threat to RSA encryption shifts from theoretical to concrete. Organisations that haven't started will be doing it in emergency mode, which costs five times as much and takes twice as long. Meanwhile, the first quantum-assisted drug candidate enters clinical trials. The headline will read: "Quantum-assisted drug enters trials." It will be underplayed by a press corps that doesn't understand why the word "assisted" matters.

By 2031 and beyond, drug discovery at major pharmaceutical companies uses quantum simulation as a standard step, not a pilot programme. Fault-tolerant quantum computing is demonstrated at small scale — proof that error correction works as theorised, though not yet scaled to solve the largest problems.

What could go wrong: hardware progress could plateau, error correction could hit unanticipated obstacles, investment could dry up. Quantum winters have happened before. They end. Add three to five years per failed assumption. What could go better: one breakthrough in qubit design or error correction could compress a decade into three years. That has happened before in computing too.

The range runs from 2028–2030 in the optimistic scenario to 2032–2035 in the conservative one. The direction is not in question. The timing is.

How to Participate (Practical Pathways)

Quantum computing needs people—not just quantum physicists. It needs engineers who can build systems that work under cryogenic constraints. Domain experts who understand what questions are worth asking. Business strategists who can tell the difference between genuine quantum advantage and quantum-flavored press releases. Policymakers who understand what they are governing. Educators who can explain qubits to people who did not study physics. The shortage is comprehensive.

Here's how to get involved:

If You're a Student:

Best preparation: Physics (foundational), Computer Science (algorithms), Electrical Engineering (hardware), Mathematics (theory), Chemistry (applications).

Essential skills: quantum mechanics, linear algebra, programming (Python dominates quantum software), statistics and probability.

Top programmes globally: MIT, Caltech, University of Waterloo, TU Delft, UC Berkeley, University of Maryland, Oxford, Yale, Harvard, University of Chicago.

Reality: Most top programmes offer full funding. Quantum PhDs are in high demand—you'll have multiple job offers before graduation.

If You're an Engineer

Learning pathway:

- Months 1–3: IBM Quantum Learning, Qiskit textbook (free, comprehensive)
- Months 4–6: Hands-on programming (IBM Quantum challenges, open-source contributions)
- Months 7–9: Domain specialization (chemistry, finance, machine learning, or optimization)
- Months 10–12: Active networking, targeted job applications

Skills that matter: Python programming (90% of quantum code), linear algebra fundamentals, quantum circuit design, classical algorithm knowledge, domain expertise in one area.

Companies hiring: IBM Quantum, Google, Microsoft, Amazon, IonQ, Rigetti, Atom Computing, plus 50+ startups.

If You're a Domain Expert

Chemists, materials scientists: Your domain knowledge matters even without deep quantum expertise initially. Learn quantum computing basics (3–6 months online), learn quantum chemistry software packages, apply to quantum chemistry companies. Roles available:

Quantum Chemistry Application Scientist

Finance professionals: Learn quantum algorithms for finance (6–12 months online), network with quantum finance startups or quantum teams at major banks. Roles available:

Quantum Finance Researcher.

Logistics and operations: Understand quantum optimization fundamentals, identify challenges in your domain, partner with quantum companies for pilots.

Roles available: Quantum Optimization Specialist.

If You Want to Start a Company

Quantum ecosystem needs: Application companies solving specific problems, software tools making quantum accessible, consulting and integration services, education and training providers.

Viable opportunities: Quantum-as-a-service for vertical industries, vertical-specific quantum software, quantum consulting for enterprises, quantum workforce development platforms.

Investment thesis: Hardware companies are capital-intensive with long timelines. Software and application companies offer better risk/reward for traditional VC. Look for teams combining quantum expertise with deep domain knowledge.

If You're in Policy or Governance

Quantum governance is severely understaffed relative to importance. Roles needed urgently: Quantum policy analysts, quantum standards developers, quantum security specialists, quantum ethics researchers.

Pathway: Develop quantum literacy through online courses, combine with existing policy/legal/ethics expertise, target government agencies, standards bodies, research institutions.

Free resources to start today, including links and a step-by-step setup guide, are in Appendix A

The Critical Point

You can participate in quantum computing revolution without PhD-level expertise. Software engineers can transition in 12 months. Domain experts in 1–2 years. Interested amateurs can gain genuine literacy in 3–6 months.

The quantum workforce shortage is severe. Demand massively exceeds supply. If you start learning now, you position yourself for careers that barely exist yet but will be in acute demand by 2028–2030.

Waiting is also a choice. It is worth being clear-eyed about what it is a choice for.

Both are choices.

The Ultimate Question

Nineteen chapters explained what quantum computing is, how it works, what it will do, and who's building it.

The more important question is not technical. It is political. Who decides how quantum technology gets used, who has access to it, and what happens to the people and nations that do not?

Who Decides

Nineteen chapters explained how quantum computing works. The more important question is who decides what we do with it.

Right now, the decisions that will shape the quantum age for the next fifty years are being made in a small number of rooms. Corporate strategy sessions at IBM, Google, Microsoft, Amazon. Government ministries in Washington, Beijing, Brussels, London, Delhi. Venture capital firms choosing which companies get funded — and therefore which capabilities get built first. University research departments setting the agenda for which questions get asked.

Most people affected by quantum computing have no seat at any of those tables. Workers whose jobs will be automated by quantum optimisation. Patients waiting for drugs that quantum simulation might design faster. Citizens whose communications are encrypted with standards quantum computers will eventually break. Nations without quantum programmes that will find themselves dependent on capabilities they cannot build.

This pattern is familiar. Nuclear energy was developed for weapons; governance came only after Hiroshima demonstrated what ungoverned looked like. The internet was built by academics and the military, governed decades after deployment. Social media was deployed at massive scale and regulated only after the damage was impossible to ignore. AI: widely adopted in 2023, governance still being negotiated.

Quantum computing is early enough that a choice still exists — not about whether quantum technology gets built, but about the priorities embedded in it. Which problems get solved first: drug discovery or military cryptography? Who gets access: institutions with $15 million R&D budgets, or anyone with a laptop and a cloud subscription? What happens to workers displaced by quantum optimisation before retraining programmes exist to absorb them?

These are political questions. They get answered politically — by who has influence over funding, regulation, and public attention.

That includes you. More than the technical complexity of the subject implies.

The Stakes—Two Futures

If the next decade goes well: quantum simulation accelerates drug discovery enough to find treatments for diseases that have resisted classical methods for decades. Materials science advances fast enough to matter for the battery and clean-energy technologies the climate requires. Quantum optimisation improves resource allocation globally in ways that reduce waste rather than simply shift costs. New industries form.

If it goes badly: encryption breaks before critical infrastructure migrates, creating a window of vulnerability for banking, healthcare, and military communications. Quantum sensors enable surveillance capabilities that outpace the privacy regulations written for cameras and microphones. Quantum optimisation displaces workers faster than retraining programmes can absorb them, in countries without safety nets built for that kind of shock. The capabilities concentrate in the handful of nations and corporations rich enough to build them, creating dependencies as deep as oil and as hard to unwind as semiconductors.

The difference between those two futures is not technical. It is governance—deliberate choice made before the crisis rather than in response to it.

The Historical Moment

Quantum computing will not solve all problems. It will not cure all diseases, reverse climate change, eliminate inequality, or end geopolitical competition.

It is a tool—extraordinarily powerful, potentially transformative, genuinely new in its capabilities—but a tool.

The question of what we choose to build with it is not answered by the technology. It is answered by us.

In 1975, personal computers were hobbyist curiosities. In 1985, useful business tools. In 1995, essential infrastructure. In 2005, ubiquitous. Quantum computing in 2026 resembles personal computers circa 1977.

In 2026 it works demonstrably, runs expensively, requires expertise, solves specific problems, and is improving at a rate that makes the trajectory visible even when the arrival dates are not. Not yet infrastructure. But clearly important and rapidly maturing.

If the trajectory holds—and there is genuine reason to believe it will—quantum utility emerges broadly around 2030, infrastructure status around 2035, ubiquity around 2040. Slower than the advocates claim. Faster than the sceptics allow. Roughly on the schedule that transformative technologies tend to follow once the foundational physics is settled.

But here's the critical difference:

When personal computers arrived, the risks were modest. Some jobs displaced. Some industries disrupted. But no existential security threats. No surveillance dystopia. No fundamental inequality amplifiers. No arms race dynamics.

Quantum computing carries all of those risks simultaneously.

Which means we can't afford to repeat the pattern of "build first, govern later, manage consequences reactively."

We need to get this right. And we're running out of time.

The Invitation

This book explained what quantum computing is and why it matters.

Now you decide what to do with that knowledge.

The invitation is not complicated. Learn quantum programming, or start a quantum company, or advocate for thoughtful policy, or teach quantum concepts to people who have not encountered them, or participate in quantum governance—or, at minimum, pay attention when the headlines arrive. Understanding what you are reading is itself a form of participation.

The future is being built right now. In laboratories and boardrooms and government offices worldwide.

You can participate. Or watch.

But you cannot claim you weren't informed.

Same technology. Different outcomes. The choice is ours.

One Last Analogy

In 1903, two bicycle mechanics from Ohio flew a powered aircraft for twelve seconds at Kitty Hawk. The longest flight that day covered 260 metres and reached an altitude of about 3 metres. Journalists who covered the story struggled to convey why anyone should care. "Flying machine," one noted, "makes uncertain leap." The New York Times reported the event. In paragraph six. Beneath an article about a snowstorm.

Sixty-six years later, humans walked on the Moon.

The gap between twelve seconds at Kitty Hawk and Neil Armstrong's footprint in lunar dust is the distance between "demonstrated possible" and "transformed civilisation." It took sixty-six years. It required two world wars, a space race, the invention of jet engines, computers, telecommunications, and the organisational capability to coordinate hundreds of thousands of people toward a single goal.

But it started with twelve seconds in North Carolina, and two people who believed they could build something that had never existed before.

Quantum computing's Kitty Hawk moments have already happened. The Wright Brothers had twelve seconds. We have demonstrated quantum error correction below the critical threshold. We have entangled photons across 7,600 kilometres of space. We have built machines colder than anything in the known universe and used them to simulate molecules that classical physics cannot model.

We don't know when the Moon landing comes. But Kitty Hawk is behind us.

The next sixty-six years will be extraordinary. Genuinely, verifiably, historically extraordinary. Not because quantum computing will solve every problem—it won't. Not because it will be easy—it isn't. But because the problems it can solve are exactly the problems we cannot afford to leave unsolved: the drugs we can't design, the materials we can't simulate, the encryption we can't secure, the climate we can't model accurately enough to act on with confidence.

Twenty chapters ago, you didn't know what a qubit was. You almost certainly didn't know quantum computers were already being used to design molecules, optimise delivery routes, and detect fraud — or that the encrypted messages you sent this morning may be sitting on someone's server waiting for a quantum computer to read them a decade from now.

Now you know. The question is what you do with that knowledge. Not eventually. Now, while the decisions are still open and the window for proactive choice has not yet closed.

> *The race nobody told you about is now the race you know about. Go tell someone. Welcome to the quantum age. Now let's decide what we build.*

References

[1] Wang, X., et al. (2016). Quantum effects in biological vision and hearing: An experimental and theoretical approach. Nature Communications, 7, 13163.

[2] Scrutton, N. S., Basran, J., & Sutcliffe, M. J. (2008). New insights into enzyme catalysis: quantum tunneling. Biochemical Society Transactions, 47, 7971–7984.

[3] Hore, P. J., & Mouritsen, H. (2016). The radical-pair mechanism of magnetoreception. Annual Review of Biophysics, 45, 299–344.

[4] Engel, G. S., et al. (2007). Evidence for wavelike energy transfer through quantum coherence in photosynthetic systems. Nature, 446, 782–786.

[5] Cao, Y., et al. (2019). Quantum chemistry in the age of quantum computing. Chemical Reviews, 119(19), 10856–10915.

[6] European logistics company (2026). Quantum route optimization deployment results. Company internal report.

[7] Financial institution fraud detection (2026). Quantum kernel improvements. IBM Quantum Network Client Success Story.

[8] Romera-Paredes, B., et al. (2024). Mathematical discoveries from programme search with large language models. Nature, 625, 468–475.

[9] IBM Quantum (2026). Quantum Network statistics and cloud platform metrics. IBM Quantum Documentation and Annual Reports.

[10] Microsoft (2024–2026). Azure Quantum Elements customer deployments and case studies. Microsoft Cloud Platform Technical Reports.

[11] Amazon Web Services (2026). Braket platform usage statistics and customer adoption metrics. AWS Quantum Computing Documentation.

[12] Finke, D. (2026). Annual quantum computing industry analysis 2026. The Quantum Computing Report. quantumcomputingreport.com.

[13] Cloud Computing Market (2026). Global revenue analysis and growth projections. Gartner Cloud Services Market Report.

[14] Nath S. Molecular mechanisms of energy transduction in cells: engineering applications and biological implications. *Adv Biochem Eng Biotechnol.* 2003;85:125–180.

Appendix A

Quantum Programming: Getting Started

For readers ready to write quantum code

Chapter 20 explained what quantum computing is, where it's going, and how to participate. This appendix is for readers who want to write quantum code today—not next year, not after a PhD, **today**.

Everything here runs on real quantum computers. Free. From your laptop. No special equipment required.

You will write two programmes.

1. The first is Grover's algorithm—quantum search, the clearest demonstration of quadratic quantum speedup on a problem everyone understands.
2. The second is QAOA—quantum optimisation, the approach already running in financial and logistics applications described in earlier chapters.

Both run on IBM's quantum hardware. Both are foundational. Both are free.

Let's begin.

SECTION 1: SETUP (5 MINUTES)

Get access to IBM Quantum:

1. Visit quantum-computing.ibm.com
2. Create free account (requires email)
3. Access: 10 minutes monthly quantum compute time (sufficient for learning)

Note: If any URL has changed since publication, search 'IBM Quantum' or 'Qiskit' directly—both are maintained by IBM and the landing pages are stable.

Install Qiskit locally (optional but recommended):

```
pip install qiskit qiskit-aer qiskit-ibm-runtime
```

Verify installation:

```
import qiskit
print(qiskit.__version__)  # Should print version number
```

Your first quantum circuit (30 seconds):

```
from qiskit import QuantumCircuit

# Create circuit with 2 qubits
qc = QuantumCircuit(2)
qc.h(0)          # Hadamard gate (superposition)
qc.cx(0, 1)      # CNOT gate (entanglement)
qc.measure_all()

print(qc)        # Displays circuit diagram
```

That's it. You're ready for real quantum programming.

SECTION 2: GROVER'S ALGORITHM (QUANTUM SEARCH)

The problem Grover solves is one everyone understands instinctively

Search an unsorted database of N items for one specific item. A classical computer checks them sequentially—N/2 checks on average, N in the worst case. Grover's algorithm finds the target in approximately √N checks through quantum amplitude amplification

Example: Database with 1 million items.

- Classical average: 500,000 checks
- Grover: ~1,000 checks

For this demonstration, we'll use N=8 (3 qubits). Small enough to run on simulators and understand, large enough to demonstrate the principle.

The Code

```
# ==============================
# GROVER'S ALGORITHM (N = 8)
# Target item = 5 (binary 101)
# ==============================

from qiskit import QuantumCircuit, QuantumRegister, ClassicalRegister, transpile
from qiskit_aer import Aer
from qiskit.visualization import plot_histogram
from math import pi, floor
import numpy as np
import matplotlib.pyplot as plt

# ------------------------------
# Problem Definition
# ------------------------------
n = 3                          # Number of qubits (2^3 = 8 items)
marked_item = 5                # Target item (binary 101)
N = 2**n

# ------------------------------
# Create Circuit
# ------------------------------
q = QuantumRegister(n, 'q')
c = ClassicalRegister(n, 'c')
qc = QuantumCircuit(q, c)
```

```
# ------------------------------
# Step 1: Initialize Superposition
# ------------------------------
qc.h(q)

# ------------------------------
# Step 2: Oracle
# ------------------------------
def oracle(circuit, qubits, target):
    # Convert target to little-endian binary
    target_bits = format(target, f'0{n}b')[::-1]

    # Apply X gates where target bit is 0
    for i, bit in enumerate(target_bits):
        if bit == '0':
            circuit.x(qubits[i])

    # Multi-controlled Z via H-MCX-H
    circuit.h(qubits[-1])
    circuit.mcx(qubits[:-1], qubits[-1])
    circuit.h(qubits[-1])

    # Uncompute X gates
    for i, bit in enumerate(target_bits):
        if bit == '0':
            circuit.x(qubits[i])
```

```
# ------------------------------
# Step 3: Diffuser
# ------------------------------
def diffuser(circuit, qubits):
    circuit.h(qubits)
    circuit.x(qubits)

    circuit.h(qubits[-1])
    circuit.mcx(qubits[:-1], qubits[-1])
    circuit.h(qubits[-1])

    circuit.x(qubits)
    circuit.h(qubits)

# ------------------------------
# Optimal Grover Iterations
# ------------------------------
iterations = floor((pi/4) * np.sqrt(N))
print(f"\nRunning {iterations} Grover iterations for N={N}\n")

# Apply Grover operator
for _ in range(iterations):
    oracle(qc, q, marked_item)
    diffuser(qc, q)
```

```
    # ------------------------------
    # Measurement
    # ------------------------------
    qc.measure(q, c)

    # ------------------------------
    # Simulation
    # ------------------------------
    backend = Aer.get_backend('qasm_simulator')
    transpiled = transpile(qc, backend)
    job = backend.run(transpiled, shots=1024)
    counts = job.result().get_counts()

    # ------------------------------
    # Display Results
    # ------------------------------
    print("Measurement Results:\n")

    for state, count in sorted(counts.items(), key=lambda x: x[1], reverse=True):
        percentage = (count / 1024) * 100
        print(f"|{state}⟩ : {count:4d} ({percentage:5.1f}%)")

    success_rate = counts.get(format(marked_item, '03b'), 0) / 1024 * 100

    print(f"\nTarget item = {marked_item} (binary {format(marked_item,'03b')})")
    print(f"Success rate = {success_rate:.2f}%")


# ------------------------------
# Plot Histogram
# ------------------------------
plot_histogram(counts)
plt.show()
```

Expected Output (Ideal Simulator)

```
Running 2 Grover iterations for N=8

Measurement results:
  |101⟩:  912 (89.1%)
  |011⟩:   45 ( 4.4%)
  |110⟩:   32 ( 3.1%)
  ...

Target was item 5 (binary: 101)
Success rate: 89.1%
```

Running on Real Quantum Hardware:

Replace the simulator with real IBM quantum computer:

```python
from qiskit_ibm_runtime import QiskitRuntimeService

# Setup (first time only)
# QiskitRuntimeService.save_account(channel="ibm_quantum", token="YOUR_TOKEN")

# Load service and select backend
service = QiskitRuntimeService(channel="ibm_quantum")
backend = service.least_busy(operational=True, simulator=False)

print(f"Running on: {backend.name}")

# Run circuit
transpiled = transpile(qc, backend, optimization_level=3)
job = backend.run(transpiled, shots=1024)
print(f"Job ID: {job.job_id()}")

# Results (may take minutes in queue)
counts = job.result().get_counts()
```

Noise on real hardware will reduce success rate to 60–80% depending on the quantum computer's current calibration.

SECTION 3: QAOA FOR OPTIMIZATION

The Problem: Portfolio optimization. You have $100,000 to invest across 4 assets. Each asset has expected return and risk. Some assets are correlated (diversification benefit). Find the optimal portfolio allocation.

Classical approach: Evaluate all combinations (2^4 = 16 for simple binary choices, exponentially more for continuous allocations). For realistic portfolios with hundreds of assets, becomes intractable.

Quantum approach (QAOA): QAOA encodes the optimisation problem as quantum states and uses quantum interference to amplify good solutions—the same mechanism Grover uses, applied to a broader class of problems.

The Code

```
# ========================================================
# QAOA PORTFOLIO OPTIMIZATION (4 Assets, Choose 2)
# ========================================================

from qiskit_optimization import QuadraticProgram
from qiskit_optimization.algorithms import MinimumEigenOptimizer
from qiskit_algorithms import QAOA
from qiskit_algorithms.optimizers import COBYLA
from qiskit.primitives import Sampler
from qiskit_aer.primitives import Sampler as AerSampler

# --------------------------------------------------------
# Define Portfolio Optimization Problem
# --------------------------------------------------------

qp = QuadraticProgram()

# Binary decision variables
qp.binary_var('A')  # Tech stock (8%)
qp.binary_var('B')  # Growth stock (12%)
qp.binary_var('C')  # Bond (6%)
qp.binary_var('D')  # Dividend stock (10%)
```

```
    # Objective:
    # Maximize return - risk interaction
    # NOTE:
    # In maximization:
    #   Negative quadratic coefficient → penalizes
    #   Positive quadratic coefficient → rewards

    qp.maximize(
        linear={'A': 8, 'B': 12, 'C': 6, 'D': 10},
        quadratic={
            ('A', 'B'): -10,  # High joint risk penalty
            ('C', 'D'): 5     # Diversification reward
        }
    )


# Constraint: Select exactly 2 assets
qp.linear_constraint(
    linear={'A': 1, 'B': 1, 'C': 1, 'D': 1},
    sense='==',
    rhs=2
)

print("Portfolio Optimization Problem:")
print(qp.prettyprint())

# ----------------------------------------------------
# Solve using QAOA
# ----------------------------------------------------

# Use Aer Sampler backend
sampler = AerSampler()

qaoa = QAOA(
    sampler=sampler,
    optimizer=COBYLA(maxiter=100),
    reps=2
)

optimizer = MinimumEigenOptimizer(qaoa)

result = optimizer.solve(qp)
```

```
# --------------------------------------------------
# Display Results
# --------------------------------------------------

print("\n" + "="*50)
print("OPTIMAL PORTFOLIO:")
print("="*50)

print(result.prettyprint())

# Interpret solution
assets = ['A', 'B', 'C', 'D']
names = ['Tech Stock', 'Growth Stock', 'Bond', 'Dividend Stock']
returns = [8, 12, 6, 10]

selected = [assets[i] for i, val in enumerate(result.x) if val == 1]
total_return = sum(returns[i] for i, val in enumerate(result.x) if val == 1)

print("\nSelected Assets:")
for asset in selected:
    idx = assets.index(asset)
    print(f"  {asset}: {names[idx]} ({returns[idx]}% return)")

print(f"\nTotal Expected Return: {total_return}%")
```

Expected Output

```
Portfolio Optimization Problem:
Maximize: 8*A + 12*B + 6*C + 10*D - 10*A*B + 5*C*D
Subject to: A + B + C + D == 2

==================================================
OPTIMAL PORTFOLIO:
==================================================
objective value: 22.0
variable values: A=0, B=1, C=0, D=1
status: SUCCESS

Selected Assets:
  B: Growth Stock (12% return)
  D: Dividend Stock (10% return)

Total Expected Return: 22%
```

What This Does:

1. Problem encoding: Translates portfolio constraints into quadratic programme
2. QAOA circuit: Creates parameterized quantum circuit encoding the problem
3. Hybrid optimization: Classical optimiser (COBYLA) adjusts quantum circuit parameters
4. Iteration: Quantum circuit samples solutions, classical optimiser improves parameters
5. Convergence: After multiple iterations, finds optimal or near-optimal solution

Scaling:

At 4 assets and 4 qubits, classical and quantum are both fast—this is a demonstration, not a deployment. QAOA's advantage emerges with 50 or more assets, where classical methods struggle with the combinatorial explosion. Current quantum hardware can handle approximately 10–20 asset problems before noise dominates the signal. That ceiling is rising.

Run on Real Hardware:

Same as Grover example—replace Aer simulator with IBM quantum backend. Note: QAOA circuits are deeper than Grover, more susceptible to noise. Results on real hardware will show more variation.

Cloud Access: Step-by-Step Walkthrough

Let's run your first quantum circuit on IBM Quantum Experience (free!):

1. Go to quantum-computing.ibm.com
2. Sign up for an IBM ID
3. Open Composer or Qiskit Lab
4. Drag gates (Hadamard, CNOT, etc.) into the circuit
5. Run it on "simulator" first
6. Then choose a real backend (like ibmq_quito)
7. Wait for your results → visualize on Bloch spheres

It's like baking cookies with Schrödinger: You can't tell if they're done until you open the oven.

Note: The Bloch sphere visualisation is available in IBM's Quantum Lab environment–if you are using the basic Composer interface, you may not see it immediately. Run the circuit first; explore visualisations as your familiarity grows

BEFORE YOU CONTINUE

This appendix showed you quantum programming basics. The code runs on real quantum hardware today. The algorithms are foundational—Grover and QAOA appear in dozens of quantum applications. But remember: quantum computing is not magic. It's physics and mathematics, engineered into computational advantage for specific problems. Not all problems. Specific ones.

Learn the tools. Understand the limitations. Build realistic applications. That's how quantum computing transitions from research curiosity to practical infrastructure.

Welcome to quantum programming.

SECTION 4: NEXT STEPS

You've now:

- Written quantum code
- Run Grover's search algorithm
- Solved optimization with QAOA
- Understand quantum advantage on specific problems

Continue learning:

Official Resources:

- Qiskit Textbook: qiskit.org/textbook (comprehensive, free)
- IBM Quantum Learning: quantum-computing.ibm.com/learn (interactive courses)
- Qiskit YouTube Channel: youtube.com/@qiskit (tutorials, lectures)

Advanced Topics:

- Variational Quantum Eigensolver (VQE) for chemistry
- Quantum Machine Learning with Pennylane
- Error mitigation techniques
- Circuit optimization strategies

Community:

- Qiskit Slack: qiskit.slack.com (active community, helpful experts)
- Stack Exchange Quantum Computing: quantumcomputing.stackexchange.com
- Quantum Computing Report: quantumcomputingreport.com (industry news)

University Courses (Free Online):

- MIT 8.370x: Quantum Information Science (edX)
- Quantum Computing: Theory to Practice (Coursera, U. Toronto)
- Architecture, Algorithms, and Protocols (TU Delft, edX)

Practice Challenges:

- IBM Quantum Challenges (periodic competitions)
- Quantum Coalition Hack (global hackathons)
- Qiskit Advocate Programme (contribute to open source)

Appendix B

Choosing a Quantum Cloud Platform

Three major platforms provide free quantum access:

IBM Quantum (quantum-computing.ibm.com)

- Hardware: Superconducting qubits
- Free tier: 10 minutes monthly
- Best for: General learning, Qiskit familiarity
- Community: Largest user base, extensive documentation

Microsoft Azure Quantum (azure.microsoft.com/quantum)

- Hardware: Multiple vendors (IonQ, Rigetti, Quantinuum)
- Free tier: $500 Azure credits
- Best for: Chemistry, materials science, enterprise integration
- Tools: Q# language, Azure integration

Amazon Braket (aws.amazon.com/braket)

- Hardware: IonQ, Rigetti, D-Wave via marketplace
- Pricing: Pay per task ($0.30–0.50 per shot)
- Best for: Optimization problems, hybrid algorithms
- Integration: Standard AWS services

Google Quantum AI (limited access but publicly available)

Quantinuum H-Series (via cloud API)

Key selection factors:

- Problem type: Simulation (IBM, IonQ), Optimization (D-Wave, QAOA)
- Programming preference: Python (IBM), Q# (Microsoft), Hybrid (Amazon)
- Integration needs: Cloud services, existing infrastructure
- Budget: Free learning vs commercial deployment

For learning quantum programming, all three platforms work well.

To Choose the Right Cloud Provider, Follow the QUANTUM Checklist:

Letter	Meaning	Question to Ask
Q	Qubit Quality	Are qubits stable (T1, T2, fidelity)?
U	Use Case Fit	Simulation, optimization, or algorithms?
A	Access	Free account or paid subscription?
N	Noise	How much decoherence do they have?
T	Tools	Does it support Qiskit, Cirq, or Q#?
U	Usability	Easy dashboard, API, documentation?
M	Market Reputation	Are they used by researchers/industry?

If you can say "yes" to 5 of 7—it's a good fit.

For most readers beginning with quantum programming, the choice of platform matters less than the choice to begin. IBM Quantum is the most documented and community-supported option for newcomers. Microsoft Azure Quantum is the strongest option for chemistry and materials science applications. Amazon Braket is the most flexible for readers already working within the AWS ecosystem. Start with IBM unless you have a specific reason to do otherwise.

Appendix C

Glossary of Quantum Terms

Term	Simple definition	Analogy
Qubit	Smallest unit of quantum info; can be 0, 1, or both	A light dimmer with infinite shades
Superposition	Being in multiple states simultaneously	You being both asleep *and* at work (quantum Mondays)
Entanglement	Two qubits act as one, even when apart	Magic dice that always land on opposite faces, no matter how far apart
Measurement	Observing forces a qubit to "choose" a state	Peeking at Schrödinger's cat
Decoherence	Loss of quantum state due to noise	Coffee getting cold mid-thought
Quantum Gate	Operation that changes qubit states	Turning knobs on a cosmic mixer
Oracle	Function that tells which solution is special	A "hot/cold" hint in a treasure hunt
Grover's Algorithm	Quantum search method with $\sqrt{N}$ speedup	Everyone hums; one voice grows louder
Shor's Algorithm	Quantum code-breaker	Cracks RSA faster than gossip spreads
QAOA	Quantum optimiser for hard decisions	A quantum approach to hard decisions—finds near-optimal solutions where classical methods get lost in the combinations
Amplitude Amplification	Boosting the probability of correct outcomes	Turning up the volume on the right answer
Quantum Advantage	When quantum outperforms best classical machines	Formerly called quantum supremacy—Winning a chess game blindfolded
Quantum Cloud	Online access to real quantum devices	Netflix for qubits

Term	Simple definition	Analogy
Bloch Sphere	A 3D representation of qubit states	The Earth map of a qubit's mood
Quantum Noise	Random errors during computation	Static in a cosmic radio
Error Correction	Technique to stabilize qubits	Putting seatbelts on excited electrons
Quantum Error Correction	Redundant qubits detect & fix errors without disturbing data.	A proofreader who watches your hands typing and corrects mistakes without ever reading the sentence.
Physical Qubit	Real hardware qubit; noisy and error-prone.	A musician in an orchestra—skilled but affected by temperature, vibrations, and nearby players.
Logical Qubit	Stable qubit built from many physical qubits via error correction.	A full orchestra section playing in unison—individual players wobble, but the combined output stays stable.
Surface Code	The leading method for error correction where qubits are arranged in a 2-D grid and neighbouring qubits continuously check each other for errors without collapsing the quantum state.	A spell-checker that never reads your text—it only checks whether neighbouring letters remain consistent.
VQE (Variational Quantum Eigensolver)	Hybrid quantum–classical method to find lowest molecular energy.	Two people solving a maze together—quantum tries thousands of paths at once, classical decides which paths to refine.
T_1 (Relaxation Time)	How long a qubit keeps its excited state before losing energy and collapsing to $\|0\rangle$—typically microseconds.	Like how long a ball thrown in the air stays up before gravity pulls it down.
T_2 (Coherence Time)	How long a qubit maintains its quantum phase relationship before noise destroys it; always $\leq T_1$.	How long you can keep a thought clear before background noise makes you lose focus.
NISQ (Noisy Intermediate-Scale Quantum)	"Noisy Intermediate-Scale Quantum": today's hardware with ~50–1,000 qubits—useful but too noisy for full error correction.	A prototype car: it runs and teaches you the future, but you wouldn't drive it across a continent yet.

Appendix D

Symbols & Representations in Quantum Computing

Symbol	Name	Simple Meaning
$\|0\rangle$	Ket zero	Classical 0, definite computational basis state
$\|1\rangle$	Ket one	Classical 1, definite computational basis state
$\|\psi\rangle$	Ket psi	Any pure quantum state
α, β	Amplitudes	Complex numbers whose magnitudes squared give probabilities
$\|+\rangle$	Plus state	$(\|0\rangle + \|1\rangle)/\sqrt{2} \rightarrow$ Equal (50/50) superposition
$\|-\rangle$	Minus state	$(\|0\rangle - \|1\rangle)/\sqrt{2}$
H	Hadamard gate	Converts $\|0\rangle$ or $\|1\rangle$ into equal superposition
X	Pauli-X (NOT)	Bit flip: $\|0\rangle \leftrightarrow \|1\rangle$
Z	Pauli-Z	Phase flip: adds −1 phase to $\|1\rangle$
Y	Pauli-Y	Bit flip + phase flip
S	S-gate	90° phase shift on $\|1\rangle$
T	T-gate	45° phase shift on $\|1\rangle$
$Rz(\theta)$	Z-rotation	Arbitrary phase rotation around Z-axis
CNOT	Controlled-NOT	Flip target if control qubit = 1
CZ	Controlled-Z	Add −1 phase if control qubit = 1
SWAP	Swap gate	Exchanges two qubits
•	Control dot	"If I'm 1, activate the gate"
$\oplus$	X-target	"Flip me" (used in CNOT diagrams)
M	Measurement	Collapse state to 0 or 1 and read result
ρ	Density matrix	Describes mixed or noisy quantum states
$Tr(\cdot)$	Trace	Sum of diagonal elements (used in probabilities & partial trace)
$\langle\psi\|$	Bra psi	Row-vector (dual) form of $\|\psi\rangle$
$\langle\psi\|\varphi\rangle$	Inner product	Overlap between two quantum states
$\otimes$	Tensor product	Combines multiple qubits/systems
$(\|00\rangle + \|11\rangle)/\sqrt{2}$	Bell state (Φ^+)	Maximally entangled two-qubit state
[[n,k,d]]	Quantum code	n physical qubits protect k logical qubits, distance d
Surface Code	Surface code	Leading practical quantum error-correction method

To You, The Reader

You just spent hours—maybe days—reading 408 pages about quantum computing. Most people won't. They'll wake up one day in 2031 and wonder what happened to the world. Now, you'll recognise the headlines. You'll understand the stakes. You're now among the relatively small number of people globally who have read past the headlines and into the substance of what is actually happening. That understanding gives you voice in decisions that will affect billions.

Now I need your help.

I spent three years on this book, but I know I got few things wrong. The field moves too fast. I missed things. Made mistakes. Explained some parts badly. Your industry might not even be in here. You see what I couldn't see. You know what I don't know. You understand contexts I missed.

So if you noticed:

- Topics that should have been covered and weren't
- Errors that need fixing
- Explanations that confused rather than clarified
- Perspectives from your country or field that are missing

I want to know. The book gets better because readers make it better. Email me your feedback and suggestions at mmbaskr@gmail.com

I read every email. Your feedback directly shapes future editions. Readers have already fixed technical errors, suggested topics I missed, and helped me make difficult concepts more accessible.

That matters more than you might realize.

One more ask.

If this book changed how you think, tell one person about it. Not everyone. Just one.

Someone who makes tech decisions at your company. A friend who asks about the future. A student figuring out their career. Someone who'd benefit from understanding what you now understand.

That's how ideas spread. Person to person. Not through ads.

And if you have five minutes, an Amazon review helps. Your honest assessment helps other people decide if this is worth their time.

Thanks for being here. For caring. For reading to the end. Keep learning. This book is a beginning, not an end.

The quantum age is arriving whether we are ready or not. You are more ready than you were 408 pages ago. Use it.

With genuine gratitude,
Dr. Mithun Manalikandy, FIE

Disclaimer

This book synthesizes decades of quantum computing research, drawing from hundreds of scientific papers, industry reports, and expert insights. The discoveries described here belong to the researchers, engineers, and theorists who advanced quantum mechanics and quantum computing from theoretical physics to practical technology.

I have organised this vast body of knowledge into a coherent narrative, explaining complex concepts through accessible analogies, and analyzing implications for technology, security, economics, and society. The scientific foundations are well-established. The interpretations and policy recommendations are my own.

Quantum computing is a rapidly evolving field. Information accurate as of early 2026 may be superseded by new developments. Timeline predictions (Chapters 19–20) represent educated estimates based on current trajectories, not guarantees. Companies, technologies, and capabilities mentioned reflect the state of the field as of early 2026. In a domain moving this quickly, some details will have changed by the time you read this. That is a feature of the subject matter, not a failure of the reporting.

This book is intended for general informational and educational purposes. It does not constitute professional advice—financial, legal, career, or otherwise. Readers considering quantum-related investments, career transitions, or policy decisions should consult appropriate professionals.

All errors, omissions, and oversimplifications are mine alone.

ILLUSTRATIONS AND VISUAL CONTENT

Illustrations, diagrams, and infographics throughout this book were created specifically for educational purposes using AI-assisted design tools and professional illustration software. Circuit diagrams follow standard quantum computing notation. Data visualizations reflect cited sources. Maps and timelines synthesize information from multiple published sources as referenced in chapter footnotes.

While every effort has been made to ensure visual accuracy, illustrations are simplified representations designed for clarity rather than technical precision. Readers requiring exact technical specifications should consult primary sources cited in references.

Dr. Mithun Manalikandy, FIE
February 2026

Index

Page numbers in italics indicate the primary discussion of a term.

M

N

P

Q

R

S

T

U

V

W

Z

oOo

www.ingramcontent.com/pod-product-compliance
Lightning Source LLC
LaVergne TN
LVHW071221220926
848486LV00038B/285